Geochemical Abundances of Selected Elements in Earth's Crust
(parts per million)

Element	Symbol	Atomic Number	Atomic Weight*	Average Continental Crust	Seawater
Aluminum	Al	13	26.98	82,300	0.01
Antimony	Sb	51	121.75	0.2	5×10^{-4}
Arsenic	As	33	74.92	1.8	0.003
Barium	Ba	56	137.34	425	0.03
Beryllium	Be	4	9.01	2.8	6×10^{-7}
Bismuth	Bi	83	208.98	0.17	2×10^{-5}
Boron	B	5	10.81	10	4.6
Bromine	Br	35	79.91	2.5	65
Cadmium	Cd	48	112.40	0.2	1×10^{-4}
Calcium	Ca	20	40.08	41,000	400
Carbon	C	6	12.01	200	28
Chlorine	Cl	17	35.45	130	19,000
Chromium	Cr	24	52.00	100	5×10^{-5}
Cobalt	Co	27	58.93	25	1×10^{-4}
Copper	Cu	29	63.54	55	0.003
Fluorine	F	9	19.00	625	1.3
Gallium	Ga	31	69.72	15	3×10^{-5}
Germanium	Ge	32	72.59	1.5	6×10^{-5}
Gold	Au	79	196.97	0.004	$<1 \times 10^{-5}$
Hydrogen	H	1	1.008	1400	108,000
Iodine	I	53	126.90	0.5	0.06
Iron	Fe	26	55.85	56,000	0.01
Lead	Pb	82	207.19	12.5	3×10^{-5}
Lithium	Li	3	6.94	20	0.17
Magnesium	Mg	12	24.31	23,000	1350
Manganese	Mn	25	54.94	950	0.002
Mercury	Hg	80	200.59	0.08	3×10^{-5}
Molybdenum	Mo	42	95.94	1.5	0.01
Nickel	Ni	28	58.71	75	0.002
Nitrogen	N	7	14.01	20	0.5
Oxygen	O	8	16.00	464,000	857,000
Palladium	Pd	46	106.4	0.01	$<1 \times 10^{-5}$
Phosphorus	P	15	30.97	1050	0.07
Platinum	Pt	78	195.09	0.005	$<1 \times 10^{-5}$
Potassium	K	19	39.10	21,000	380
Rubidium	Rb	37	85.47	90	0.12
Selenium	Se	34	78.96	0.05	4×10^{-4}
Silicon	Si	14	28.09	282,000	3.0
Silver	Ag	47	107.87	0.07	4×10^{-5}
Sodium	Na	11	22.99	24,000	10,500
Sulfur	S	16	32.06	260	885
Thorium	Th	90	232.04	9.6	5×10^{-5}
Tin	Sn	50	118.69	2.0	8×10^{-4}
Titanium	Ti	22	47.90	5700	0.001
Tungsten	W	74	183.85	1.5	1×10^{-4}
Uranium	U	92	238.03	2.7	0.003
Vanadium	V	23	50.94	135	0.002
Zinc	Zn	30	65.37	70	0.01

*Source of atomic weights: 177–178, 1965. Based on atom〔 〕hemistry, *Compte. Rendu.*, XXIII Conf., pp. 〔 〕nal places.

D1465505

RESOURCES OF THE EARTH

Origin, Use, and Environmental Impact

Second Edition

JAMES R. CRAIG
*Virginia Polytechnic Institute
and State University*

DAVID J. VAUGHAN
The University of Manchester

BRIAN J. SKINNER
Yale University

Prentice Hall
Upper Saddle River, New Jersey 07458

Library of Congress Cataloging-in-Publication Data

Craig, James R.
    Resources of the earth / James R. Craig, David J. Vaughan, Brian
J. Skinner.
      p.   cm.
    Includes bibliographical references and index.
    ISBN 0-13-457029-4
    1. Natural resources.   2. Environmental policy.   I. Vaughan,
David J.   II. Skinner, Brian J.   III. Title.
HC21.C72   1996
333.7—dc20   96-33650
                                 CIP

Editorial Director: Tim Bozik
Editor-in-Chief: Paul F. Corey
Acquisitions Editor: Robert A. McConnin
Director of Production/Manufacturing: David W. Riccardi
Special Projects Manager: Barbara A. Murray
Total Concept Coordinator: Kimberly P. Karpovich
Production Supervision/Composition: BookMasters, Inc.
Illustrations: Patrice Van Acker; Charles Pelletreau; Peter Ticola; BookMasters, Inc.
Insert Composition: Karen Stephens
Art Director/Cover Design: Jayne Conte
Cover Photo Credits: Background photo: Bingham Canyon Mine in Utah (Courtesy of Kennecott Corporation)
                      Top Inset: Gulf War 1990-1991 (Jonas Jordan, U.S. Army Corps of Engineers)
                      Middle Inset: Gold nugget from the Whitehall Mine, Spotsylvania County,
                           Virginia (Smithsonian Institution, NMNH #995251)
                      Bottom Inset: Glen Canyon Dam, Colorado (Courtesy of U.S. Bureau of Land
                         Management)

© 1988, 1996 by Prentice-Hall, Inc.
Simon & Schuster / A Viacom Company
Upper Saddle River, New Jersey 07458

ISBN 0-13-457029-4

Printed in the United States of America
10 9 8 7 6 5 4 3 2

Prentice-Hall International (UK) Limited, *London*
Prentice-Hall of Australia Pty. Limited, *Sydney*
Prentice-Hall Canada Inc., *Toronto*
Prentice-Hall Hispanoamericana, S.A., *Mexico*
Prentice-Hall of India Private Limited, *New Delhi*
Prentice-Hall of Japan, Inc., *Tokyo*
Simon & Schuster Asia Pte. Ltd., *Singapore*
Editora Prentice-Hall do Brasil, Ltda., *Rio de Janeiro*

in one great stride and in 1854 he commenced work on a monster ship of 22,500 tons which he planned to be six times the size of anything afloat. Built on the Thames at Millwall, the *Great Eastern* was to remain the largest ship built in the world for forty-one years. When she was launched in January 1858, after refusing to leave her slip for three months despite Brunel's efforts, the new giant was in fact 18,915 tons and 680 feet in length; her nearest rival was Cunard's *Scotia* at 3,850 tons.

Marine engineering in 1858 was hard pressed to provide engines sufficiently advanced for the size of the ship and Brunel had to rely on sail, paddle and screw, using all three together and singly. The combination was not successful and even though the huge hull was only fitted out for the modest capacity of

300 passengers and 6,000 tons of cargo, the *Great Eastern* never attracted enough trade to sustain the costs of the large crew and engine-room staff required for her five boiler rooms and two engines. Her great advantage was her huge fuel capacity (12,000 tons) which gave her a working range of 7,000 miles at seven knots. The costs of the massive ship were too much for the stock markets and she bankrupted her first owners and killed Brunel by the overwork and strain caused by the technical and economic problems of his huge creation. He died on 15 September 1859, just eight days after the *Great Eastern* had sailed on her trials.

Ignoring the fact that the *Great Eastern*'s design made her unsuitable for the Atlantic, the new owners sailed the ship on her maiden voyage to New York on 17 June 1860, with a paltry thirty-

eight paying passengers and crew of 418! She was greeted by crowds of enthusiastic New Yorkers who swarmed all over her. But she never made money and by 1864 had been sold for cable-laying operations. It was then she came into her own. Between 1867 and 1874, the *Great Eastern* laid five transatlantic cables and one between Suez and Bombay. Then she was laid up for eleven years and finally scrapped in Liverpool, beginning in 1888. The scrapping took three years to complete.

The *City of New York* had twin screws and was also one of the first liners to dispense with sail, so reliable had engines become by the 1880s. She and her sister *City of Paris* were elegant ships and must be numbered among the most beautiful ships of all time. They dwarfed the current Cunard fliers *Umbria* and *Etruria*,

In 1901, a ship was finally built to exceed the Great Eastern *in every way; this was the* White Star Celtic *(below), commissioned for the Liverpool-New York route.*

Holland America Line poster of 1880, showing the W. A. Scholten.

which were still only single-screw. In 1893, Cunard replied with the *Campania* and the *Lucania* (12,950 tons), and these two sisters regained the trade for Cunard, crossing in five and a half days, well inside a week and three times as fast as the old *Britannia* of 1840.

But in 1897, British complacency was shattered by a brilliant newcomer from Germany, the *Kaiser Wilhelm der Grosse*, the new flagship of North German Lloyd. This ship can be recognized as the first true modern luxury ocean liner. She was large (14,349 tons and 627 feet in length),

powerful, and her accommodation met the standards now being demanded by the prosperous American businessmen who emerged from forty years of unchecked economic growth in America. She soon received a Marconi wireless telegraph installation, being the first Atlantic liner to have one. The glamor of this one ship brought so much business to North German Lloyd, that in 1898 the company carried 28 per cent of all passengers to arrive in New York. The company followed with three similar ships, all slightly larger in size, all with

four funnels as in the *Kaiser Wilhelm der Grosse*, and Hamburg-Amerika added the *Deutschland*, a similar ship with four funnels like her predecessors.

Germany ruled on the Atlantic and, to add to the troubles of an embattled Cunard Line, the White Star Line of Liverpool introduced their *Oceanic* in 1899, the first ship to exceed the *Great Eastern* in length. They followed up with the *Celtic*, first to exceed the *Great Eastern* in all ways, and added the *Cedric*, *Baltic* and *Adriatic* in the early years of the new century. There was much Ameri-

The crew of the Augusta Victoria, a Hamburg-Amerika liner of 1889.

MITTELMEER-FAHRT DER

Decksmannschaft d

UGUSTA VICTORIA" 1891.

Augusta Victoria".

can money in the White Star Line, although it was British registered and manned. J. Pierpont Morgan, the archetypal American tycoon, had intervened in shipping and by 1900 he had control of most of the big American and British companies on the Atlantic, with the exception of Cunard.

Alarmed by the threatened extinction of the nation's premier flag-carrier, the British Government intervened and did so at a time when another technical advance was to change dramatically the established engine designs used in liners.

All the nineteenth-century liners were driven by reciprocating steam engines, in which steam expanded inside a cylinder to drive a piston which in turn rotated the propeller shaft. As ships grew in size, the demand for more power caused the designer to increase greatly the size of his engines. These massive forgings, driving round at 80 r.p.m. in the *Kaiser Wilhelm der Grosse* to produce 28,000 i.h.p., created massive vibration which affected the whole ship. The largest set of reciprocating engines ever fitted went into the North German Lloyd liner *Kronprinzessin Cecilie* in 1904. But it was obvious that the reciprocating engine had reached the end of its possible development.

A new source of power was required, and it came from the ideas of Charles Parsons, a British engineer who invented a steam turbine in 1884.

Parsons' invention was a brilliant success. Turned away by a disbelieving Admiralty, he placed his turbine in a small launch, suitably called *Turbinia*, and demonstrated his little vessel's superiority in speed by a convincing and unauthorized intrusion into the lines of assembled warships in the 1897 Spithead Review of the British Fleet. Their Lordships sent their latest torpedo-boat destroyer to chase Parsons' boat away and the little *Turbinia* left the naval vessel standing. The turbine had the advantage of much greater power-weight ratio, lower maintenance costs and vibration-free running compared to the huge reciprocating engines necessary to drive

a ship like the *Kaiser Wilhelm der Grosse* across the Atlantic at an average speed of 20 knots or so. Immediate experiments in other ships were carried out and the first transatlantic liner to use turbines, the Allan Line vessel *Virginian* came out in 1904: another step forward.

Already in January 1902 Morgan had convened a conference in New York of British and American ship owners. The result of this international conference was an agreement whereby the entire White Star Line and its fleet were sold to the American group together with the Atlantic Transport and Leyland companies. The acquisition by a foreign financial organization of a prestige British shipping company caused a sensation among a generation far more susceptible to patriotic fervor than the people of today. In consequence an Admiralty committee under the chairmanship of Lord Camperdown got down to work immediately. The result was that the Admiralty entered into a contract with the Cunard company to build two ships, capable of a minimum speed of not less than 24·5 knots in moderate weather. For this Cunard was to receive an annual subsidy of £150,000 and a loan of £2,500,000.

Cunard was already building two large intermediate liners, the *Caronia* and the *Carmania*. *Carmania* was hastily re-engined with turbine machinery and the critical and, as it proved, key decision was made to engine the two new ships with turbine machinery also. Contracts for the two vessels were placed respectively on the Tyne and Clyde, with orders going to Swan Hunter and to the Clydebank yard of John Brown and Company. The two ships were to become the immortal *Mauretania* and her sister *Lusitania*.

The *Mauretania* of 1907 is arguably the most famous ship ever built. At 31,938 tons she was far and away the largest ship in the world and dwarfed her German rivals. She measured 790 feet and she clipped almost a whole day off the Atlantic speed record. Her best passage before the First World War was four days, ten hours

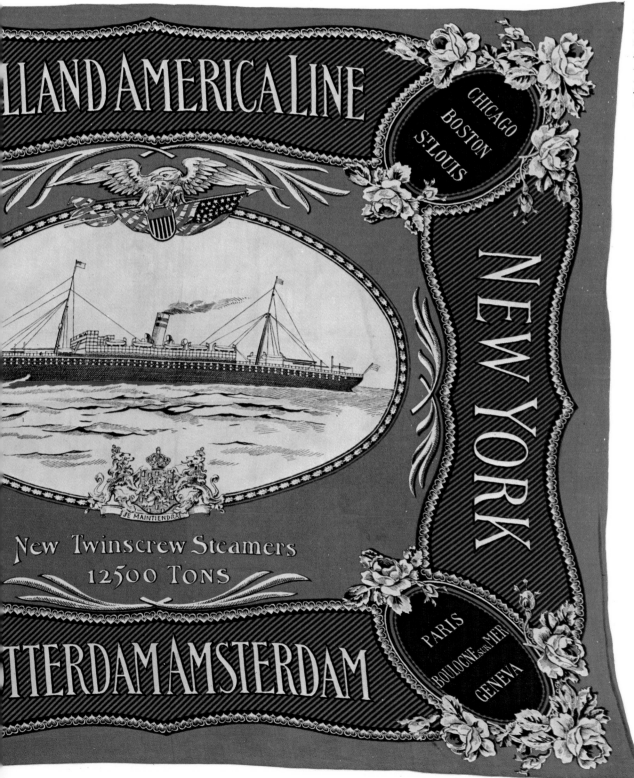

HOLLAND AMERICA LINE

NEW YORK

CHICAGO
BOSTON
St. LOUIS

New Twinscrew Steamers
12500 TONS

JE MAINTIENDRAI

ROTTERDAM AMSTERDAM

PARIS
BOULOGNE SUR MER
GENEVA

*A handkerchief
commemorating the
introduction of twin-
screw steamers by the
Holland America
Line in 1900; the ship
shown is the Potsdam
(12,000 tons).*

*The building of the
Imperator in 1913 in
Hamburg; the rudder
of the 52,000-ton ship
towered over the
shipyard.*

*Parsons' Turbinia,
the pioneering steam-
turbine launch.
Below:
the Turbinia seen
next to the elegant
form of the
Mauretania, which
led the British effort
to dominate the
north Atlantic
passenger trade for
many years.*

and fifty-one minutes at an average speed of 26·06 knots to recover the record, and the *Mauretania* held it for twenty-two years, a brilliant achievement!

After the *Mauretania*, ships could only get bigger. The White Star produced *Olympic* (1911) and her tragic sister *Titanic* (1912), both at 46,000 tons, and followed with the 48,000-ton *Britannic* in 1914.

Albert Ballin, now in charge of Hamburg-Amerika, replied with three even larger ships in the years 1913 and 1914. The *Imperator* (52,117 tons), *Vaterland* (54,282 tons) and *Bismarck* (56,551 tons) were to remain the world's largest ships for twenty-five years, but after the war

they were ceded as war reparations and sailed as the *Berengaria* (Cunard), the *Leviathan* (United States Lines), and the *Majestic* (White Star). After the First World War the economic depression caused the Atlantic passenger trade to decline, but it did not halt the production of superliners. Germany launched the *Bremen* (54,725 tons) and the *Europa* (49,746 tons) in 1928, and the Italians produced the *Rex* (51,062 tons) and the *Conte di Savoia* (48,502 tons) in 1932.

The thirties saw the apotheosis of the superliner. One hundred years on from the early voyages of Cunard ships, Cunard, after a merger with the ailing White Star, produced the huge *Queen Mary* to

challenge the equally large *Normandie*, star of the French Line. The latter entered service in May 1935, and the *Queen Mary* followed a year later. Both ships were overtaken in size by the *Queen Elizabeth* (83,763 tons) launched in Scotland on 27 September 1938.

In just a century liners had grown from the 1,300 tons of the *Great Western* to the 83,000 tons of the *Queen Elizabeth*. The journey to New York had been reduced from over two weeks to less than four days and the big ships had become symbols of national superiority and pride. At the time, few people realized that the age of the superliner had only thirty more years to run.

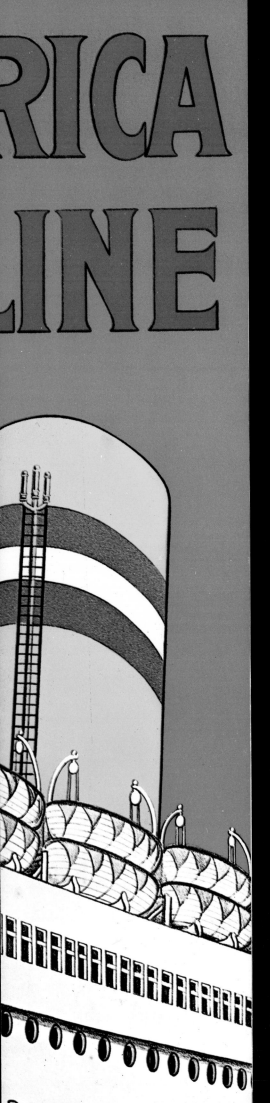

CHAPTER TWO

FLYING THE FLAG

The story of the great lines

A poster detail of the
Statendam III *(1927)*.

As the 19th century faded, the infant American republic came of age and took her place as one of the world's most powerful and economically prosperous nations. The long march westward was over, the bloody agony of Civil War forty years behind, and the railways now carried the wealth of the industrial north to the four corners of the American continent. The unprecedented economic expansion in the United States between 1865 and 1900 caused an immediate shortage of the one prime commodity that America did not possess – people. At the same time, such European social disasters as the famine in Ireland, rebellion and civil strife in the Balkans and Eastern Europe, allied to abject poverty in the Latin countries, particularly Italy, caused a stream of emigrants to turn their eyes westward and sail for the New World.

This huge emigrant trade brought prosperity to the shipping lines on the north Atlantic and, in the opposite direction, a demand rose from prosperous Americans for luxury travel to Europe. On the north Atlantic, therefore, a multimillion dollar passenger trade grew up and the big shipping companies which engaged in this traffic prospered and grew.

At the same time as the industrial advances made in America, the British Empire also prospered, and great shipping lines grew up on the routes that went out through the Mediterranean and the Suez Canal to India and the East, or dipped across the Indian Ocean to Australia and New Zealand and across the far reaches of the Pacific to the coast of South America. British imperial expansion in South Africa also resulted in a passenger line reaching down through the South Atlantic to Cape Town and beyond.

In the latter years of the 19th century, Germany and France also began acquiring colonial empires. German interests in East and Central Africa and her great commercial activities in South America led to the formation of German shipping lines running to the coasts of those

The rapid growth of the great north Atlantic passenger lines during the 19th century owed much to the large numbers of Europeans who wished to emigrate to the North American continent. Here, a group of eastern European emigrants pose on board a ship of the Holland America Line, while others (inset) wait outside the Cunard office in Trieste.

countries. France also took a great interest in the South American trade and both the major European powers joined Britain in a struggle for dominance on the north Atlantic.

From profits gained in the early years of the trade, when steamship design advanced every year, there grew up in Europe a series of great shipping lines whose prosperity added much to the strength of each national economy and most European governments were careful to protect and enhance the reputation of their big flag carriers. A disaster to a shipping company could have dire consequences on the stock markets, leading to depression and unemployment, such was the power of the companies which had developed and grown over the years. The oldest of the Atlantic lines was the Cunard Steamship Company. As we have already seen, this famous establishment was founded by Samuel Cunard in 1839 as the British and North American Royal Mail Steam Packet Company.

Sam Cunard, as we have seen, learned his business skills in the company founded in Halifax by his father Abraham. Young Sam went to Halifax Grammar School and later he worked for three years in Boston, in the office of a ship broker. By the time he was twenty-five he was already operating ships and the firm of Abraham Cunard & Son had some forty vessels. It was then only a short step to the founding of the Cunard Steamship Line. A friend described Cunard as 'a bright, tight little man with keen eyes, firm lips and happy manners'. Not only did he run his steamship company, but he was also the Commissioner of Lighthouses, a member of the Nova Scotia parliament and involved in several charities. His intervention in the British business scene was equally successful and in his later career he was the friend of several prime ministers. At his death, the boy whose early beginnings on the streets of Halifax had led to the command of a vast shipping empire, had been created a baronet.

Cunard and his partners achieved immediate success and their mail con-

CUNARD R·M·S·AQUITANIA

tract was renewed in 1846. Four years later they faced serious competition when the American Collins Line began sailings from New York. Although the Cunard ships now sailed from New York instead of Boston, they were still outclassed by their larger and faster American competitors on the north Atlantic.

The Collins Line was probably one of the most unfortunate enterprises ever to embark passengers for the Atlantic trade. E. K. Collins founded the line in 1846. He thought that America should try to occupy the same place in the world of steam that she had built for herself with her clippers, which had deservedly won a world-wide reputation. Collins therefore asked Congress to provide a subsidy for a steamship line operating between New York and Liverpool under the American flag. His proposals won Congressional approval and the result was that Collins built four splendid wooden steamers called *Atlantic, Arctic, Baltic* and *Pacific,* all of 2,860 tons and 282 feet in length.

The new vessels were a great advance on the existing Cunard ships. They were the first Atlantic liners to have smoking rooms specially set apart; other innovations included bathrooms and barber shops. The *Atlantic* also appears to have been the first liner to fit an engine-room telegraph for communication between the bridge and the engine-room far below. This was a great advance on the voice pipes, which had been very inefficient, and most captains still stationed a boy within hearing of the bridge to shout orders down through the engine-room skylight. Steam heating also was an attractive innovation and hot water was supplied even to the steerage quarters. Nevertheless it was to be another twenty years before the French Line introduced running water into all cabins, the first company to do so. This ended the early morning ritual on all liners when bedroom stewards could be seen standing in line along the leeside rail ceremoniously emptying chamber pots!

The Collins Line commenced operation in April 1849 and the *Atlantic* and her sisters attracted numerous passengers from Cunard. The Collins ships, however, were never a paying proposition and Sam Cunard himself, pointing out the expensive equipment of the American ships, described their efforts as 'breaking Cunard windows with sovereigns'. It was in September 1854 that the Collins Line received the first blow which led to its extinction. The *Arctic* collided with a French steamer *Vesta* off the Newfoundland banks in fog and sank with a loss of 322 lives including Collins' wife and his son and daughter. Less than two years later he suffered another blow when the *Pacific* sailed from Liverpool on 29 June 1856 and vanished at sea. Undeterred, Collins produced a larger ship in 1857. The *Adriatic* measured 3,670 tons and was a great advance on her sisters. She cost so much that the United States government grew restive and finally withdrew its subsidy; the Collins Line collapsed in January 1858. It had made a bold attempt to challenge Cunard and no major American Atlantic line emerged for over sixty years until the United States Lines was founded in the 1920s.

The Cunard fortunes continued to prosper. The company encountered no real competition until the end of the century when, as we have seen, Germany produced a series of splendid liners. By 1900 Cunard fortunes were changing.

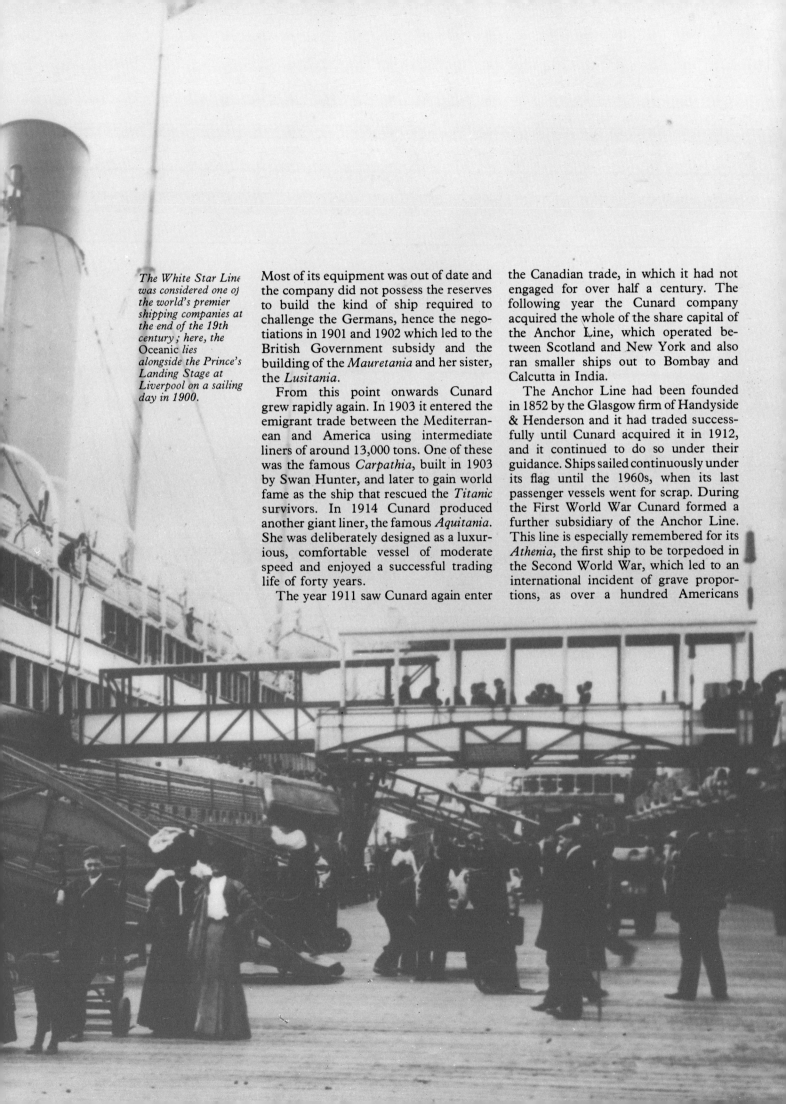

The White Star Line was considered one of the world's premier shipping companies at the end of the 19th century; here, the Oceanic *lies alongside the Prince's Landing Stage at Liverpool on a sailing day in 1900.*

Most of its equipment was out of date and the company did not possess the reserves to build the kind of ship required to challenge the Germans, hence the negotiations in 1901 and 1902 which led to the British Government subsidy and the building of the *Mauretania* and her sister, the *Lusitania*.

From this point onwards Cunard grew rapidly again. In 1903 it entered the emigrant trade between the Mediterranean and America using intermediate liners of around 13,000 tons. One of these was the famous *Carpathia*, built in 1903 by Swan Hunter, and later to gain world fame as the ship that rescued the *Titanic* survivors. In 1914 Cunard produced another giant liner, the famous *Aquitania*. She was deliberately designed as a luxurious, comfortable vessel of moderate speed and enjoyed a successful trading life of forty years.

The year 1911 saw Cunard again enter

the Canadian trade, in which it had not engaged for over half a century. The following year the Cunard company acquired the whole of the share capital of the Anchor Line, which operated between Scotland and New York and also ran smaller ships out to Bombay and Calcutta in India.

The Anchor Line had been founded in 1852 by the Glasgow firm of Handyside & Henderson and it had traded successfully until Cunard acquired it in 1912, and it continued to do so under their guidance. Ships sailed continuously under its flag until the 1960s, when its last passenger vessels went for scrap. During the First World War Cunard formed a further subsidiary of the Anchor Line. This line is especially remembered for its *Athenia*, the first ship to be torpedoed in the Second World War, which led to an international incident of grave proportions, as over a hundred Americans perished. Cunard suffered heavy losses during the First World War, including the *Lusitania*, but by the mid-twenties this tonnage had been replaced and later in the same decade the Cunard board were planning a replacement for its ageing 'Big Three' liners, the *Mauretania*, *Aquitania* and *Berengaria*. In 1930 it laid down on the Clyde an 80,000-ton giant, to be built by John Brown and Company at their Yard No. 534. The ship was no sooner started than the company was hit by the economic depression of those years. Falling passenger figures on the Atlantic and a loss of trade caused work on the new liner to be held up and thousands of Clydeside workers laid off. The future looked grim until the British Government intervened and, in exchange for a subsidy to continue the construction of the new ship, ordered a merger of Cunard and White Star, its principal British competitor.

The White Star Line dated back to 1869 when a company known as the Oceanic Steam Navigation Company Limited was founded in Liverpool by Thomas Ismay. This was a steamship company and an offshoot of the White Star fleet of sailing clippers. The white star on a red pennant which had been flown by the ships of the old organization was now adopted as the house flag of the new company. It always traded under the title White Star Line, but its official name remained Oceanic Steam Navigation.

Thomas Ismay stands out as the leading British shipowner of the last part of the 19th century. Always bold in his ideas and aggressive in his commercial policies, he had turned the White Star Line into the most powerful British shipping company by the end of the century. His passenger ships were always larger than their predecessors and he traded not only to America but also to Australia and South Africa. The Australian service began in 1883 and the South African service followed fifteen years later.

Ismay launched the *Oceanic* in 1899 and then proceeded to an even larger ship, the *Celtic* of 1901. But Ismay never

The White Star Line's Celtic *was launched in 1901 to consolidate the line's pre-eminent position on the north Atlantic route. It was, however, only half the size of the huge* Olympic *(1901) and* Titanic *(1911), seen here at Harland and Wolff's yard in Belfast, where all the major White Star ships were built.*

52

R.M.S. Celtic
Length, 697 ft.
Breadth, 75 ft.
Tonnage, 20,904

saw the *Celtic* enter service. He died in 1900 and two years later his White Star Line was acquired by the Morgan combine and became a member of the International Mercantile Marine Company. J. Pierpont Morgan was an American financier who had built up an industrial empire in steel. He had been interested in the Atlantic shipping trade as early as 1893 and set out to get a monopoly on the north Atlantic. In addition to the White Star Line, Morgan acquired the Leyland and Atlantic Transport Lines, and he reached agreement for cooperation with the two large German companies, the Hamburg-Amerika line and North German Lloyd.

With American capital behind it and using the skills of the great Harland and Wolff shipbuilding yard at Belfast (the White Star Line never went to any other shipyard) the new combine immediately began plans for two giant liners, which were to become the *Olympic* and her tragic sister the *Titanic*. The ships of the line remained under the British flag and were operated under British crews; and in 1907 the line moved its headquarters from Liverpool to Southampton. The great days of the Atlantic Ferry were about to begin. The *Olympic* was introduced in 1911 and was greeted by the press as the 'World's Wonder Ship', which indeed she was. There followed the *Titanic* a year later, and then tragedy struck. The new liner on her maiden voyage collided with an iceberg off the Newfoundland Banks on the night of 14–15 April 1912 and sank with the loss of 1,500 lives. Among the survivors was Bruce Ismay, managing director of the White Star Line and son of its founder. Ismay was much criticized for entering a lifeboat when many passengers were unable to be rescued and he never recovered from the effects of the disaster. Within a year he had resigned his post and retired from public life to become a virtual recluse.

After the *Titanic* disaster the White Star was never the same again. It lost another large ship, the *Britannic* of 1914, during the First World War and, al-

though it received the huge *Majestic* (the ex-German *Bismarck*) as war reparations in 1920, its fortunes declined seriously and in 1927 it returned to British ownership, only to go through yet another series of financial crises. These difficulties led to the cancellation of a proposed new liner, the third *Oceanic*, which was to have been a 60,000-ton motor vessel. The keel for the third *Oceanic* was actually laid at Belfast in 1928, but work proceeded slowly and the project was eventually cancelled. This would have been a remarkable ship. The proposals included a three-funnelled liner with twenty-four diesel engines, each driving an electric generator, which in turn drove four screws.

By 1933 the worldwide slump had hit the White Star Line especially hard and only four ships were running to New York and a fifth to Canada. At the insistence of the British Government the White Star was amalgamated with Cunard.

Cunard White Star Line Limited, as the merged companies were called, came into being on 1 January 1934. Most of the White Star tonnage was sold and the line concentrated on producing *534* (the *Queen Mary*) by 1936 and following up with the *Queen Elizabeth* whose entry into service was planned for 1940, the centennial of the line. In the meantime Cunard produced a third large liner, the second *Mauretania* of 1938.

The outbreak of the Second World War caused the cancellation of the *Queen Elizabeth*'s maiden voyage and it was not until 1947 that the Cunard dream of two large, 80,000-ton liners operating a weekly service became reality. In the years immediately after the war the service prospered and the company repaid in full the government loan which had been the means of building the *Queens*. But by the 1950s, aviation had begun to dominate the North Atlantic passenger trade and one by one the great Cunarders were withdrawn from service. Nevertheless, again with a government subsidy, Cunard built the *Queen Elizabeth 2* which entered service in 1969 and today remains the world's largest passenger ship. Cunard itself was taken over by a finance company, Trafalgar House, in 1971 and this company breathed new life into the old shipping line. Today Cunard concentrates on its cargo services and cruising is the main activity of its three passenger vessels. Nevertheless, each summer for several months, *QE2* provides a north Atlantic service and so the old Cunard tradition sails on.

Throughout its long history, the Cunard Line faced fierce competition from its two great German competitors for much of the time. In the year 1847, as we have seen, the Hamburg-Amerika Line had been formed to establish a line of sailing vessels to New York. By 1856 it had introduced its first steamship and from that time onwards Hamburg-Amerika vessels maintained a passenger service from Europe to New York which was only interrupted by the two World Wars. During that time Hamburg-Amerika produced some of the world's most outstanding ships and much of their success was due to Albert Ballin who became chairman of the company in 1880.

When it came to running passenger ships Albert Ballin was a genius. He had a flare for knowing exactly what his passengers required and he was capable of extraordinary energy. He spoke fluent English and was in constant touch with his competitors in the English shipping companies. When Ballin travelled on one of his ships, his notebook was never out of his hand. He scribbled away on such details as the print on the passenger list, the toast at breakfast, dirty sheets, the quality of the playing cards in the smoking room, and the hard pillow in his cabin. No detail was too small for Ballin to consider and the quality of the service on his liners improved accordingly.

Ballin, more than any other shipowner, foresaw the possibilities of taking luxury to sea, and he was the first magnate to employ specialist interior decorators and chefs de cuisine aboard his ships. He enjoyed the patronage and friendship of Kaiser Wilhelm II and possessed a flair

The interior décor of the first-class dining room in one of Ballin's ships, the Kaiserin Auguste Victoria, *seen below passing the Statue of Liberty.*

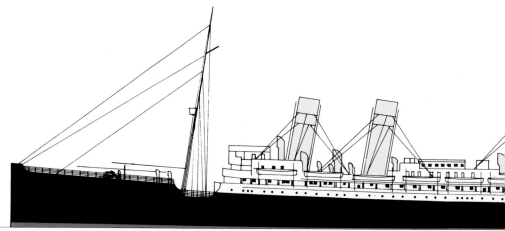

for handling the capricious young monarch which even Bismarck envied.

He died tragically by his own hand in 1918, unable to face the loss of his great ships which the approaching defeat of Germany would make inevitable.

Ballin began to build large express steamers in 1887 and he did so in direct reply to North German Lloyd which had been attracting many passengers from his company, since his ships were smaller and slower. Ballin's new ships came into service in 1889 and their success was immediate, for his *Columbia* and *Augusta Victoria* were among the finest ships of their day.

By the year 1900 the Hamburg-Amerika Line had become a powerful organization with passenger freight services to most parts of the world. It was then that Ballin produced his *Deutschland* of 16,500 tons. He already possessed the largest fleet of merchant ships owned by any one company and now he produced the fastest ship ever built to that date. The *Deutschland*, although a splendid advertisement for the company when she captured the north Atlantic record, proved very expensive to run and convinced Ballin that the future of his passenger services lay in large ships rather than in fast ones. Having obtained an agreement with the Morgan organization to guarantee his continued independence, and not wishing to challenge Cunard and the *Mauretania*, Ballin adopted a policy of large, comfortable and relatively slow vessels, with the result that his *Imperator* of 1912 was a 52,226-ton monster, driven by turbine engines, the first to be fitted in any large German ship. *Imperator* was followed in 1914 by her larger sister *Vaterland*. Both these ships had marvellous accommodation of which Ballin had every reason to be proud and he proposed to add a third great ship, the *Bismarck* of 56,000 tons when, in the summer of 1914, the outbreak of war intervened.

The First World War practically annihilated German shipping and Hamburg-Amerika suffered with the rest. When it re-entered the race in the 1920s

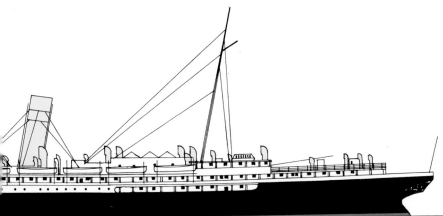

Ballin's attempts to compete with the North German Lloyd in the race for the fastest crossing of the Atlantic were never completely successful. Although the Deutschland *of 1900* did capture the Blue Riband on her maiden voyage, she suffered from constant mechanical trouble throughout her career. The main dining room of the Deutschland *again shows the love of ostentation and luxury characteristic of all Ballin's ships.*

Kaiserlich Deutsche Post.

Norddeutscher Lloyd, Bremen.

Norddeutscher Lloyd

Directe regelmäßige Postdampfschifffahrt

nach Nord=Amerika.   nach Süd=Amerika.

zwischen

Bremen und Newyork,

Baltimore, Neworleans, Havana, Brasilien und La Plata.

Der **Norddeutsche Lloyd** in Bremen ist die größte deutsche Dampfschifffahrts-Gesellschaft. Sie besteht bereits seit dem Jahre 1857 und erfreut sich einer immer zunehmenden Beliebtheit bei dem reisenden Publikum, was am deutlichsten daraus hervorgeht, daß bis Ende vergangenen Jahres 681,387 Personen mit diesen Schiffen gereist sind. Die Gesellschaft hat soviel Vertrauen zu der soliden Bauart der Dampfer und der Tüchtigkeit ihrer Capitaine, daß sie nicht allein die ganze Assecuranz auf die Schiffe, sondern außerdem noch bedeutende Versicherungen auf die zur Versendung kommenden Güter und Passagierseffecten übernimmt. Die Schiffe werden daher, um Schaden, den der Lloyd selbst zu tragen hätte, zu vermeiden, stets sorgfältig untersucht und im besten Stande erhalten, und den Capitainen ist die äußerste Vorsicht zur Pflicht gemacht. Dadurch wird den Passagieren die größte Sicherheit gewährleistet. Die Dampfschiffe des Norddeutschen Lloyd, welche bereits 2764 Reisen glücklich über den Ocean gemacht haben, fahren:

I. Zwischen Bremen und Newyork:

Abfahrt von Bremen: Sonntags, von Southampton: Dienstags.

	von Bremen:	von Newyork:		von Bremen:	von Newyork:		von Bremen:	von Southampton: Dienstags		von Bremen:	von Newyork:		von Bremen:	von Newyork:
D. Mosel			D. Weser	1. Nov.		D. Main	2. Nov.	22. Nov.	D. Donau	23. Nov.	13. Decbr.	D. Main	14. Decbr.	3. Jan. M.
D. Oder		25. Octbr.	D. Rhein	8. Nov.		D. Mosel	9. Nov.	29. Nov.	D. Rhein	30. Nov.	20. Decbr.	D. Mosel	21. Decbr.	10. Jan. M.
D. Donau		1. Nov.	D. Neckar	15. Nov.		D. Werder	16. Nov.	6. Dec.	D. Neckar	7. Decbr.	27. Decbr.	D. Oder	28. Decbr.	17. Jan. M.

II. Zwischen Bremen und Baltimore direct:

	von Bremen:	von Baltimore:		von Bremen:	von Baltimore:		von Bremen:	von Baltimore:		von Bremen:	von Baltimore:
D. Baltimore		16. Octbr.	D. Hohenzollern	13. Nov.		D. Baltimore	12. Nov.		D. Leipzig		10. Dec.
D. Ohio		31. Octbr.	D. Leipzig	5. Nov.	27. Nov.	D. Ohio	26. Nov.	18. Decbr.	D. Baltimore	24. Dec.	15. Jan. M.

III. Zwischen Bremen und Neworleans,

event. via Hâvre und Havana:

von Bremen:

D. Hannover 19. November.

IV. Zwischen Bremen, Bahia, Rio de Janeiro und Santos,

via Antwerpen und Lissabon:

	von Bremen:	von Antwerpen:	von Lissabon:
D. Graf Bismarck		29. October.	4. 5. Nov.
D. America	25. Novbr.	29. Novbr.	4. 5. Decbr.
D. Berlin	25. Decbr.	29. Decbr.	4. 5. Jan. 1880.

V. Zwischen Bremen, Montevideo und Buenos Ayres,

via Antwerpen und Bordeaux nach den La Plata-Staaten:

	von Bremen:	von Antwerpen:	von Bordeaux:	von Buenos Ayres:
D. Salier				25. Sept.
D. Hohenstaufen				25. October.
D. Habsburg				25. Novbr.
*D. Salier	10. Novbr.	14. Novbr.	19. Novbr.	25. Decbr.
D. Hohenstaufen	10. Decbr.	14. Decbr.	19. Decbr.	25. Jan. 1880.

*D. Salier 10. November läuft **Madeira** an.

VI. Zwischen Bremen und London:

Abfahrt von Bremerhaven: Jeden Sonntag und Mittwoch, Morgens.

" London: Sonnabend und Mittwoch, Mittags.

VII. Zwischen Bremen und Hull:

Abfahrt von Bremerhaven: Jeden Montag, Morgens.

" Hull: Sonnabend, Abends.

NB. Extradampfer werden expedirt, wenn der Verkehr es erfordert.

Bremen, im October 1879.

Die Direction des Norddeutschen Lloyd in Bremen.

An early poster for the North German Lloyd service between Bremen and New York. The company was founded by a wealthy merchant of Bremen, H. Meier. The brilliant Kaiser Wilhelm der Grosse *of 1897, seen here at New York, was sometimes called 'Rolling Billy'; she ensured North German Lloyd first place on the North Atlantic until Cunard produced the* Mauretania *of 1907.*

Ballin was dead (he died before the end of the war, heartbroken at the ruin of his life's work) and a policy of small and intermediate liners was adopted. The first ship was appropriately called *Albert Ballin* and she made her maiden voyage to New York in 1923. Less than half the size of her predecessors, the new vessel was 20,000 tons. She was later joined by three sister ships. By 1929, Hamburg-Amerika was once again the world's largest ship owner, but it never built an Atlantic giant again and, when history repeated itself in 1939 and the company again lost all its tonnage, the rebuilt, post-war organization concentrated largely on freight. Hamburg-Amerika eventually amalgamated with the North German Lloyd to become the present-day company known as Hapag-Lloyd.

The North German Lloyd was founded in 1857 by an influential merchant from Bremen called H. Meier. Its first steamer was built in Scotland on the Clyde, a single-screw vessel of 2,674 tons which took the name of its home port of Bremen. The service prospered and by 1887 the Lloyd was running ships to the Caribbean, Brazil and to Argentina. With the increasing size of its ships, it was forced to move down to the Lower Weser at Bremerhaven, and at this time it also acquired its own Hoboken terminal at New York.

The North German Lloyd suffered its own share of disaster in the 19th century, something which only Cunard of the pioneer companies seems to have avoided. In 1875 the *Deutschland*, groping her way down the Channel in a fog, went aground on the Goodwin Sands and was a total loss. Later came the wrecks of the *Condor*, *Hansa* and *Mosel*, and in 1892 the *Eider* went aground on the Isle of Wight.

Worse was to follow. The *Elbe* was sunk in a collision in the North Sea with 335 lives lost, and in June 1900 a disastrous fire destroyed the company's piers in New York and seriously damaged the *Bremen*, *Main* and *Saale*.

But the company's flagship, the brilliant *Kaiser Wilhelm der Grosse* of 1897 was towed clear. It was this ship, together with her sisters, which gave the North German Lloyd domination of the North Atlantic until 1907, when Cunard produced the *Mauretania*.

In 1902 the North German Lloyd formed a partnership with Morgan, but retained its autonomy. The company was now led by Dr. Wiegand. Unlike his opposite number at Hamburg-Amerika, Wiegand firmly believed in fast ships and was not enamored of very large liners. He therefore built medium-sized ships; his largest ship prior to the First World

NORDDEUTSCHER LLOYD

Lloyd-Gesellschaftsreisen.

D. *Schnelldampfer Columbus*

Vom 29 Juli bis 6 August 1933

Abfahrtshafen *Bremerhafen*

Name *Elisabeth Haase*

Zimmer № *887*

BREMEN

The North German Lloyd liner Columbus (32,500 tons) (her first-class saloon is seen here) of 1923 was converted into a turbine ship in 1929 to increase her speed, so that she could act as a running mate to the crack Bremen and Europa. Right: a poster of 1907, celebrating fifty years of North German Lloyd.

War was the *George Washington* of 24,000 tons which ran between Bremen and New York. Until the arrival of the *Columbus* after the war, *George Washington* remained the largest of the Lloyd steamers. Dr. Wiegand died in March 1909 and was succeeded by Phillip Heineken. Like all the large German companies, the North German Lloyd lost all its ships during the First World War, and the early twenties found it rebuilding its fortunes. After the building of *Columbus*, a fine passenger liner of 32,500 tons, a number of small ships were produced until the Lloyd dramatically re-entered the race for the

Atlantic record in 1929 with the two superliners *Europa* and *Bremen*.

These wonderful ships were prominent on the north Atlantic right up to the outbreak of war in 1939. They held the record until 1933 when the Italians wrested the crown from them. Nevertheless the *Bremen* and *Europa* remain among the top six or seven crack liners ever built.

The *Bremen* was destroyed by fire during the war, while the *Europa* was ceded to France to become the *Liberté* in 1945. Once again the North German Lloyd was obliged to rebuild a passenger fleet. This it did by using old Swedish

tonnage and the *Berlin* became its first liner after the war. Then it acquired the large French vessel *Pasteur* which had been built originally for the South American routes. Renamed *Bremen* she went into service between Germany and America. Later still, another *Europa* was acquired in the form of the Swedish motor-liner *Kungsolm* and this ship remains in operation, mainly on cruising, up to the present time. Meanwhile, the North German Lloyd amalgamated with Hamburg-Amerika in 1971.

We have noted that North German Lloyd and the Hamburg-Amerika Line

1857 · 1907

Dem
Norddeutschen Lloyd
zu seinem 50 jährigen
Bestehen
F. Schichau.

BREMEN
The German superliner

Like all the large German companies, North German Lloyd lost its major liners after the First World War. In 1929, the Lloyd dramatically re-entered the race for the Blue Riband with the *Europa* and the *Bremen*, which must be rated among the top six or seven liners ever built. The long, sleek lines of the *Bremen* gave her an entirely new and modern profile, while her interiors reflected the revolutionary concepts of German design of the 1920s.

Reichspresident von Hindenburg at the launching of the Bremen *in 1929 at Bremerhaven.*

The massive bulk of the Bremen *on the slipway (left); finally entering the water (right).*

The catapult aircraft of the Bremen *was used for carrying express mail to New York; the aircraft would be launched when the liner was in flying distance of the shore.*

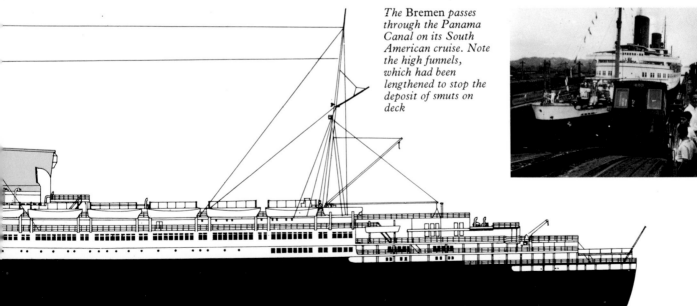

The Bremen *passes through the Panama Canal on its South American cruise. Note the high funnels, which had been lengthened to stop the deposit of smuts on deck*

The Compagnie Générale Transatlantique, more commonly known as the French Line, was the first company to introduce proper interior plumbing to ocean travel, in the Normandie *of 1883, seen here at Le Havre. The line was founded in 1855 by the brothers Emile and Isaac Pereire and became one of the strongest international lines, as this poster shows.*

were both involved in the South America passenger trade, but the most important of the German passenger lines to South America was that known as the Hamburg South American Line. This company was founded in 1871 and by 1914 it was the largest on the south Atlantic. In that same year it had produced the *Cap Trafalgar*, but this ship lasted only a year and was sent to the bottom by the British liner *Carmania* in a vicious battle off Trinidada in the autumn of 1914. The Hamburg South American Line operated the largest ship on the south Atlantic up to the outbreak of the war in 1939. This was the

Cap Arcona of 27,560 tons, an impressive three-funnel liner which was destroyed by bombing during the war.

Many British companies operated to South America. These included the Royal Mail Line, which commissioned, notably, the 27,000-ton *Andes*, so popular with British cruising enthusiasts after the Second World War, and the Pacific Steam Navigation Company which traded down the west coast of South America to Chile. The Blue Star Line, the Blue Funnel Line, the Booth Line and the Nelson Line were all prominent British companies trading to South America. The prize for building the largest ship ever to sail to South American waters, however, must go to the French company, the Compagnie Sud-Atlantique.

The South Atlantic company occupied much the same position on the south route as the French Line did on the north. Its mail boats maintained a regular service from Bordeaux to Buenos Aires and in 1930 they produced a 34,000-ton liner, *L'Atlantique*, which was far away the largest ship ever to enter the River Plate. Unfortunately, when she was only three years old, *L'Atlantique* caught fire outside Le Havre while on a voyage to be refitted, and became a total loss. The company built the *Pasteur* to replace her but, after the Second World War, the trade declined and the *Pasteur* never sailed on the route for which she was designed.

Not far behind the German and British lines in prominence was the great French organization, the Compagnie Générale Transatlantique. Known to travellers in Britain and the United States as the French Line, the company had been founded in 1855 under the brothers Emile and Isaac Pereire. They received encouragement from the French government, which was alarmed at the weakness of French merchant shipping which had been demonstrated during the Crimean War of 1854. By 1861 the company was operating a fleet of twenty-seven vessels and, although it suffered a setback during the Franco-Prussian War of 1870, by 1885 it was one of the strongest lines on the

*Below : the French
Line's* Paris
*(34,500 tons) of 1919
putting out to sea.
Right : the flair and
elegance of French
interior design seen in
the grand staircase of
the* Paris.

*Three favorites of
the north Atlantic
run at Le Havre in
1935 ; left to right :
the* Paris, *the* Ile de
France *and the*
Normandie.

north Atlantic. The *Normandie* was the first liner to be equipped with interior plumbing, and the company followed this with a series of good quality, medium-size vessels right into the early years of the 20th century. In 1912 the company commissioned its first large liner, the *France*, whose four funnels and luxurious décor made her an immediate favorite with the transatlantic passenger. She was the first large turbine liner to be built in France, and her British port of call was always Plymouth. After the First World War the French Line introduced the *Paris* of 1919 and followed her with one of the all-time favorites of the North Atlantic, the splendid *Ile de France*. The *Ile* was a magnificent ship and gained a deserved reputation for excellence. Veteran travellers were prepared to wait a week in order to sail on her and her name found its way even into the popular songs of the day.

The crowning achievement of this great line came in the mid-1930s, when the famous Blue Riband winner *Normandie*, second only in size to the *Queen Elizabeth* and possibly the most outstanding passenger liner ever, was put into service. The *Normandie* possessed many innovations and her tragic loss by

*Poster by
Cassandre advertising
the French Line
service to New York.
Below : the
Normandie at the
quayside at Le
Havre ; from her
maiden voyage in
1935 to her tragic end
by fire in 1942, she
was thought by many
to be the finest ship
afloat.*

accidental fire in New York in 1942 must be rated among the great social and economic disasters of our time.

When the war ended in 1945 the French Line rebuilt the *Ile de France* and took over the ex-German *Europa*, which became the *Liberté*. The latter acted as their flagship until they produced a worthy successor to the *Normandie* in the *France* of 1961, a splendid vessel which was taken out of service in 1974. Meanwhile, the French Line has withdrawn its passenger service to Central and South America and the Far East and now largely operates ferries in the Mediterranean and freight-liner services.

It is a curious fact that, in the early years of the 20th century, while much American capital was invested in passenger shipping, no big line emerged under the American flag. J. P. Morgan, shrewd businessman that he was, always seemed content to operate through European lines, and did not seek to offend patriotic sentiment by transferring his British interests to the American register. The cheaper operating costs of British and German ships was also an influential factor in the argument.

Although there was no big American line on the north Atlantic, in other seas the Stars and Stripes flew from the mast-heads of several proud shipping companies. Typical of these was the Matson Navigation Company of San Francisco. Founded by Captain William Matson, who operated a sailing fleet between Hawaii and the American mainland, it began to run regular services in 1908 with the *Lurline*, a vessel distinguished by having her engines placed aft, the first passenger liner to do so. The *Lurline*, 413 feet in length, was followed by several similar vessels, all with blue-topped buff funnels carrying the Matson 'M'. The fleet changed its hull colours from brown to brilliant white in 1927 with the arrival of the 17,000-ton *Malolo*, a twin-funnelled ocean patriarch which is still on the register in 1977 as the Greek cruise liner *Queen Frederica*.

In the course of its long career on the service to Hawaii and on across the Pacific to Polynesia and Australia, the Matson Line absorbed several west-coast competitors. Today, in addition to a fleet of black-hulled freighters, Matson continues its Hawaii services with the small but luxurious liners *Mariposa* and *Monterey*.

Two other prominent pre-First World War American flag carriers were the Dollar Line (later to become the American President Line) on the north Pacific, and Moore, McCormack which operated down to Rio and the River Plate from east-coast ports. But not until the First World War was over did the Americans return to the north Atlantic run, when the United States Lines was founded in 1921.

At the end of the war, the United States Government found itself with a large fleet of ex-German liners which included two previous holders of the Blue Riband.

Most of these ships had been used as troopers in the years since 1917 and they all required costly refits to make them ready for passenger service again. This forlorn fleet included the *Amerika*, the *George Washington*, the *Kaiser Wilhelm II*, the *Kronprinz Wilhelm* and the monstrous *Vaterland*, now renamed *Leviathan*. The ships were under the control of the United States Shipping Board, a state agency, and it was obvious that no private owner had the funds available to bring such a vast collection of ageing liners into commercial service again. In any other country, they would have been quietly sent away for scrap, but this was the America of the twenties which had yet to experience the Depression years, and

The Pennsylvania
built in 1873 for the
Inman (later
American) Line.

Return of the Washington Volunteers from Manila
on the 'Pennsylvania' Oct 9th 1899

Known affectionately
as the 'Levi Nathan',
the Leviathan was in
fact the reconditioned
German Vaterland,
which had been
handed over to the
United States after
the First World War
as war reparation.
She sailed under the
control of the United
States Shipping
Board, a state
agency.

someone had the notion that now was the time for the United States to have its own luxury fleet on the Atlantic. So with the help of a large vote of funds from Congress, the United States Lines was founded, which took over a number of the larger liners, including the *Leviathan*. The remaining ships were laid up for use as troop transports in any future war.

Reconditioning the *Leviathan* as a passenger liner was a huge task and she was not ready until 1923. But with unlimited finance and with a design team under the brilliant naval architect William Gibbs (he later designed the *United States*), the *Leviathan* emerged an almost new ship, capable on trials of 27 knots, and her owners expected big things of her when she was put into service between New York and Southampton. The big ship was advertised as the 'World's Largest', an honor she achieved by use of the American measurement rules, by which she was judged to be 59,957 tons. The paying public was not informed that, by the same rules, the White Star's *Majestic* would weigh in at over 61,000!

The *Leviathan* was joined in service by the *George Washington* and the *America*, but the line found it almost impossible to fill its ships and it never made a profit. The reasons were not hard to find. These were the Prohibition years in the States and the provisions of the Prohibition Act extended to American ships at sea. Even foreign vessels had to padlock their bar stores at the twelve-mile limit. Given the choice, regular travellers avoided 'dry' ships and American ships suffered accordingly. Then in the mid-twenties the United States emigration laws restricted entry into the country and further trade losses followed.

By 1929, the American Government had seen too much of its money vanish up the *Leviathan*'s two smoke stacks (the third was a dummy), and the subsidy was withdrawn. Control of the line passed to P. W. Chapman who had plans for two 45,000-ton sisters to replace *Leviathan*, but the international financial depression intervened and the line changed hands again late in 1930 when it was bought by the International Mercantile Group. The *Leviathan* was laid up and sold for scrap in 1938, and the United States Lines introduced the smaller but profitable liners *Manhattan* and *Washington*, 24,000-ton sisters which became very popular, especially after

Prohibition. This allowed the building of the largest American ship to date, the *America* of 1939, 34,000 tons and named by Eleanor Roosevelt. Today the *America* is still in service as the Greek-owned *Australis*.

The Second World War taught the United States Navy much about the operation of large troopships. Americans were impressed by the impact on Allied strategy of the two Cunard *Queens* which carried a large part of the American army to Europe for the Overlord operation, which resulted in the successful invasion of Europe. National prestige also demanded that an American challenger to the *Queens* should be built. American pride never quite got over the fact that the two big sisters were British designed and owned; indeed, many GIs crossed the Atlantic in the *Queens* fully convinced that they sailed in American ships under the American flag.

When the United States Lines received a $48 millions subsidy in 1950 to produce a 50,000-ton superliner, then, naval strategy played a large part in the design which William Gibbs and his partners produced. The new ship was fast (state security dictated that her speed should

The United States *of 1952, designed by the brilliant naval architect William Gibbs, was almost certainly the fastest major liner ever built, although her speed still remains a military secret. Although by far the larger ship, her profile was very similar to that of the* America.

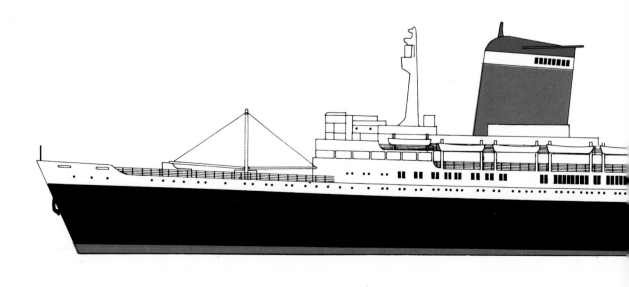

remain a secret) and she proved it by taking the Atlantic record by almost four clear knots in June 1952. This flier was the *United States*, William Gibbs' masterpiece and the fastest passenger liner ever built.

The *United States* and *America* operated together until 1965 when falling trade forced the sale of the smaller ship to the Greek Chandris Lines. They refitted her at Piraeus and in 1966 she joined their eastwards round-the-world service from Europe. Supported by a subsidy, the *United States* sailed on, until financial support ended in 1969, when the line's passenger services were withdrawn. Today the company is a highly successful operator of container express freighters between America and Europe.

Although the major share of passenger trade between Europe and the American continent passed through United States ports, there has always been a sizeable traffic between Europe and Canada, and mention must be made here of the great Anglo-Canadian concern, the Canadian Pacific Railway Company, and its subsidiary, Canadian Pacific Steamships Ltd.

The first transcontinental train from Montreal reached the Pacific coast in

The United States Lines' America of 1939 was the largest American ship at that date; she is still in service today as the Greek-owned Australis.

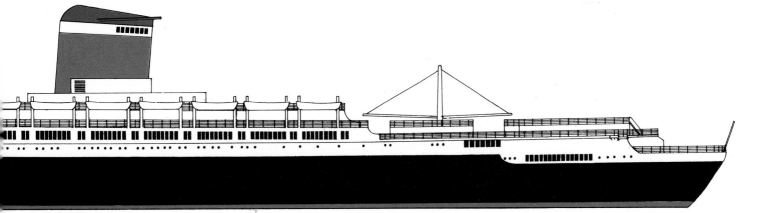

The Canadian Pacific Empress of Britain makes her stately progress up the St. Lawrence River at Quebec. Le Château Frontenac, a Canadian Pacific Hotel, dominates the skyline.

R.M.S. Victorian. (Allan Line) 12000 Tons, Turbine.

Before the First World War, the Allan Line provided competition for Canadian Pacific with the two liners, Victorian and Virginian. The two lines finally amalgamated in 1915.

July 1886, and the directors of the railway were immediately interested in a steamship line to trade across the Pacific to Yokohama. This service required fast vessels of suitable size, and to build them Government aid was needed. This came in 1889 in the shape of a mail contract from the British and Canadian authorities and three fast steamers were promptly ordered from Vickers at Barrow. These were the first of the famous *Empresses*, whose white hulls and buff funnels were a familiar sight on the north Pacific until the Second World War. The 5,900-ton *Empress of India* and her sisters, *Empress of Japan* and *Empress of China*, had clipper bows and looked very attractive. They offered a service on which a through first-class ticket from Liverpool to Yokohama cost but £68 in 1891, although the north Atlantic leg was operated by the White Star Line.

It was not until 1903 that the Canadian Pacific started an Atlantic service of its own, and this it did by purchasing fifteen ships of the Beaver Line, then owned and operated by Elder Dempster, which traded mainly to West Africa from Liverpool. Canadian Pacific added the first Atlantic *Empresses*, the *Empress of Britain* and the *Empress of Ireland*, in 1906 and this pair took on the established vessels on the route, the pioneer turbine liners *Victorian* and *Virginian* of the Allan Line.

Famous in its day but now long extinct, the Allan Line had been formed in Glasgow in 1854 for involvement in the Canadian trade and it had experienced many of the misfortunes associated with trading down the St. Lawrence River to Montreal. Between 1857 and 1865, due to insufficient navigation lights, ice and frequent fog, the line lost nine ships and a total of 650 lives. Despite keen competition from the Dominion Line and the Beaver Line, Sir Hugh Allan persevered and by the turn of the century his company was well-established. Always pro-

gressive, the Allan Line had been first with steel ships in 1879, and now in the 20th century it was first to have turbine-driven liners. With a sea speed of 18 knots and space for 1,650 passengers, the *Victorian* and *Virginian* proved successful in service until the Canadian Pacific challenged them with the larger *Empresses*. The two lines amalgamated in 1915.

In the meantime the Canadian Pacific had grown and by 1914 it was a major carrier of emigrants and freight. In that year, it suffered a serious blow when the *Empress of Ireland* was in collision in fog in the St. Lawrence and sank in twenty minutes with a death toll of 1,053 lives.

After the First World War, Canadian Pacific continued to run successfully on both oceans. New liners were built for both services and in 1931, the giant new *Empress of Britain* of 42,350 tons was introduced. The largest ship ever to trade to Canada, she had all the luxury attractions of the New York liners. Her 24

knots meant that a traveller by way of Montreal could get to Chicago more quickly than by way of New York. In the off-season the big ship did an annual round-the-world cruise and her loss in action in 1940 was a sad blow to countless travellers who remembered her with something approaching awe. She certainly created a majestic impression with her huge funnels, which were the largest ever fitted to a ship.

The Second World War wrought havoc with the Canadian Pacific fleet and it did not return to the north Pacific when peace was restored in 1945. All its Pacific vessels save one were lost in the war, and the survivor, the *Empress of Scotland* of 1930, was transferred to the Atlantic, where the services were only a skeleton of those provided before the war. In the Pacific the line formed an airline and relied solely on air transport for its service to Japan.

The *Empress of Scotland* had started life in 1930 as the *Empress of Japan*, but her name was changed for political reasons in 1942. She was joined on the Canadian Pacific post-war services by three new *Empresses*, the 25,500-ton sisters *Empress of Britain* (1956) and *Empress of England* (1957) and their slightly larger sister *Empress of Canada* (1961). All three were designed for cruising as an off-season alternative to the scheduled summer run, but the service faltered in the sixties due to competition from the airlines and all four ships were sold. The *Empress of Scotland* went to German buyers in 1958 to become the new Hamburg-Atlantic Company's *Hanseatic*, while *Empress of Britain* became the Greek Line flagship *Queen Anna Maria*. After a spell of cruising, the remaining Canadian Pacific liners were scrapped and the company today runs freight services only, except for its Canadian coastal services.

Two other European nations, Holland and Italy, were prominent on the Atlantic during the reign of the superliners. The Netherlands American Steam Navigation Company, later to gain international ac-

77

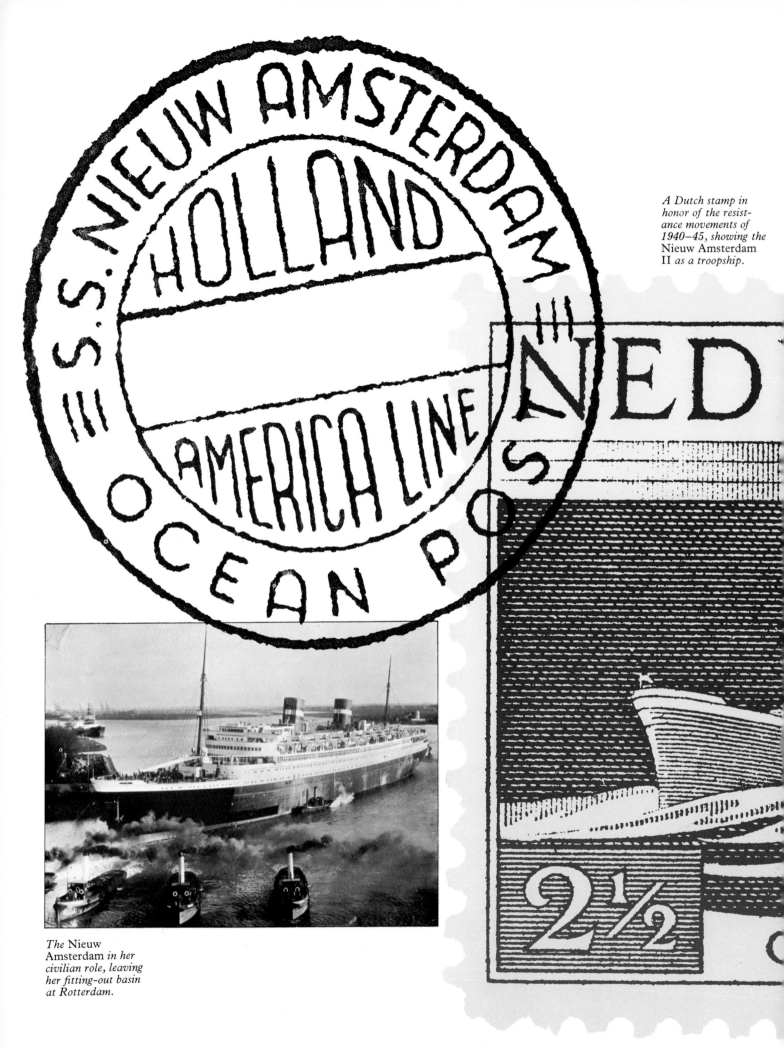

A Dutch stamp in honor of the resistance movements of 1940–45, showing the Nieuw Amsterdam II *as a troopship.*

The Nieuw Amsterdam *in her civilian role, leaving her fitting-out basin at Rotterdam.*

claim as the Holland America Line, was founded in Rotterdam in 1855. For their first ships they went to Scottish builders, and services to New York were started with the *Rotterdam* and *Maas*. These were small screw steamers of 1,700 tons, which were about the largest that the port of Rotterdam could accept in those days before the building of the New Waterway from the Hook of Holland. Indeed, for some years of its early life, the Dutch company had to sail its larger liners from the deep-water port of Amsterdam.

These early years were ones of struggle and disaster. The wreck of the steamer *Edam* in 1882 was followed in quick succession by the losses of the *Amsterdam*, *Rotterdam* and *Maasdam*, and these had to be replaced by second-hand ships from Britain. But by 1900 the line had become well-established and two years later it negotiated a shrewd deal with Morgan and his associates, whereby Holland America remained in Dutch control but got vast financial aid from Morgan's Belfast allies, the shipbuilders Harland and Wolff.

In the years that followed, Harland and Wolff produced a series of new liners for Holland America, which included the 17,000-ton *Nieuw Amsterdam* of 1906 and the 24,000-ton *Rotterdam* of 1908; this remained the largest Dutch ship up to 1914. The deal with Morgan had turned out well for the Dutch, and in early 1911 Harland and Wolff began the construction of their largest yet, the 35,000-ton *Statendam*, which was scheduled to enter service in 1914. But other events in that portentous year intruded upon the plans of the Dutch company and the great *Statendam* was destined never to see Holland. She was requisitioned by the British while still incomplete, renamed *Justicia* and placed under Cunard management. While sailing as a troopship, she was torpedoed off the coast of Ulster on 25 July 1918 and sank the following day.

That loss apart, the days of war were highly profitable for the neutral Dutchmen, and after the conflict they were able

MICHELANGELO
Italia Line luxury of the sixties

More than any other great shipping line, the Italia line long kept faith with the great traditions of North Atlantic passenger shipping. In 1965, the glistening white sisters, the *Michelangelo* and *Raffaello*, were introduced to the Genoa–New York route. Such were the standards of service and comfort on these liners, that they were busier than those of any other nation during the 1960s.

A bas-relief of Michelangelo in the first-class foyer of the liner.

The bows of the Michelangelo sweep majestically upwards, as the liner awaits her release from the slipway. Inset : the blessing of the Michelangelo.

Now painted in gleaming white, the Michelangelo leaves Genoa on her maiden voyage to North America.

Identical sisters at berth in Genoa: the Michelangelo *and* Raffaello.

The Michelangelo *brought a feeling of cruise-ship relaxation to the north Atlantic passenger trade in the 1960s.*

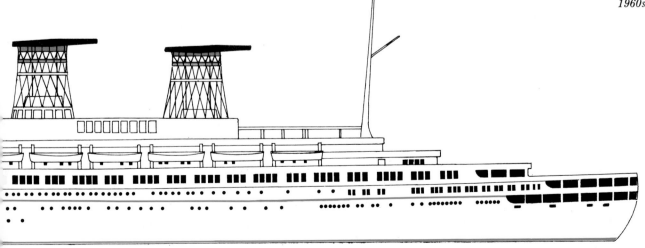

to expand their fleet with such famous ships as the beautiful three-stacker *Statendam* of 1929, still recognized as a classic example of naval architecture. She was followed by another favorite of the regular traveller, the 36,300-ton *Nieuw Amsterdam* of 1938, which became so popular with the Dutch people that an attempt to sell her for scrap in 1968 led to a public outcry, and after a refit she carried on for another five years until February 1974, when she finally went to breakers in Taiwan.

By a strange quirk of fate, the third *Statendam* was also a war loss, being destroyed during the bombing of Rotterdam in May 1940. But after the war, the *Nieuw Amsterdam* returned to service, to be joined later by yet another *Statendam* and the splendid *Rotterdam* in 1959. These last two remain in service, and although they are fully engaged on cruising, they are a living reminder of the heyday of one of the world's greatest steamship companies.

Among the many attributes of the Italian nation there must be numbered a fine sense of drama and a refined skill in the creation of beautiful things.

Both qualities were demonstrated in 1932 when the two largest Italian passenger steamship companies joined forces to

form the Societa 'Italia' di Navigazione, soon to be known the world over as the Italia Line. The partners were the old Navigazione Generale Italiana (known as the N.G.I.) and the Lloyd Sabaudo; these two later absorbed a third concern, the Cosulich Line. The N.G.I. was already operating large motor vessels, such as the 30,400-ton *Augustus* of 1927, which remains the largest diesel-powered liner ever built.

Some influence in the merger is attributed to Mussolini, who was anxious that Italian prestige be boosted by possession of an Atlantic flier and the new line did not let him down. The N.G.I. had already laid down the great *Rex,* 51,000 tons, and the Italia Line brought her into service in September 1932. Her maiden voyage was a chapter of technical mishaps, but in June 1933 she captured the Blue Riband for Italy. In company with her sister, the slightly smaller *Conte di Savoia,* the *Rex* was among the finest liners operating between the wars. She continued to operate the ferry after her British and German competitors had gone to war in 1939. Both ships had the clean lines characteristic of Italian naval architecture and both were tragically destroyed by air attack during the war.

The year 1945 found the Italia company with most of its fleet either destroyed in the war or taken over by the Allies. The Italians therefore proceeded to build a post-war fleet which exceeded all other passenger lines in the number and quality of its new vessels. First to emerge was the *Giulio Cesare,* 27,078 tons, of 1951 and her sister, the second *Augustus.* There followed another pair of lovely ships, the *Cristoforo Columbo* of 1954 and her ill-fated sister *Andrea Doria,* lost in collision with the Swedish liner *Stockholm* in fog on 25 July 1956. The *Andrea Doria* was replaced by an equally good-looking vessel, the *Leonardo da Vinci* of 1960, somewhat larger at 33,000 tons. As we shall see later, all these Italian post-war liners were luxuriously fitted out and they carried some of the best-trained crews on the Atlantic.

This luxury and service were carried to their highest refinement in the 45,000-ton sisters *Michelangelo* and *Raffaello* of 1965. These glistening white ships represented the ultimate in passenger liner design and for a while they benefited from the preference of southern Europeans for sea travel. Consequently the Italian liners in the sixties were busier than those of any other nation. But even this revival of the trade was short-lived

and by 1975 the Italia Line had withdrawn all passenger services.

Because the great spaces of the Atlantic saw the rise of the passenger liner to its Golden Age and then witnessed its long decline, it is inevitable that the larger part of this account should be devoted to ships that travelled the Atlantic routes. But other great journeys were made from European ports, often equalling the grey northern routes for danger and the need for courage and sheer physical endurance. Before the coming of the steamship and the introduction of planned schedules, sailing vessels only left port when a full load of passengers had been obtained for the trip and the traveller might find himself committed to anything up to six months confinement in a small ship on indifferent food while the vessel might equally well be bowling along in the gales of the Roarin' Forties or stuck becalmed in the staggering heat of the tropics.

Any of these conditions could be met and have to be endured on the high routes of empire to India or on to the Far East

and down under to Australia and New Zealand. Many passenger shipping companies rose and declined on these routes as the years went by, but of those that survive into the 1970s, the two major lines are both British-owned – the Union Castle and the P. & O. Only they are now left to challenge the timeless business skill of the Greeks and the vast state-aided apparat of the passenger services of the Soviet Union.

The Union Castle Mail Steamship Company was formed in February 1900 by the merger of the two largest lines trading to the British colonies in South Africa. The Union Line dated back to 1853 and its great rival, the Castle Line, had been founded by Sir Donald Currie in 1872. The *Walmer Castle*, which sailed for Table Bay in May of that year, was the forerunner of the great fleet of Castle liners to follow. At the turn of the century, the crack Union flier *Scot* and the *Dunottar Castle* were the leading contenders in the race to the Cape and the *Scot*'s record of slightly over fourteen days, which she

set in 1893, was to stand for forty-three years until broken by the motor liner *Stirling Castle* in 1936. But such rivalry was expensive and so the merger was carried through. The colors of the Union Castle were – and are – distinctive. The hull is painted lavender-grey with white upperworks and funnels are pale crimson with black tops. All the ships of the merged line received 'Castle' names.

The Union Castle was unique among shipping companies in that it was the only big company that did not lose a single major ship in the First World War. Following the war, larger ships were built and a round-Africa service started in 1922. In 1926 the company introduced the first of a long series of large motor vessels, the *Carnarvon Castle*. This form of propulsion was used to engine all its ships up to the Second World War, culminating in the 27,000-ton *Capetown Castle* of 1938.

After the war, new tonnage was ordered, but in 1956 Union Castle was taken over by the British and Commonwealth Ship-

84

Sometimes called the 'Cunard of the East', the P. & O. Line dominated the trade to India and Australia during the heyday of the British Empire. Above: the Orsova (1909) seen passing under Sydney Harbour Bridge. Right: the Orcades of 1948, one of the finest liners in the post-war P. & O. fleet.

The modern P. & O. liner Oriana *(42,000 tons) berthed at Southampton.*

ping Company, the London-based owners of the Clan Line. The new owners produced the largest vessels ever to sail on routes to southern Africa: the *Windsor Castle* of 1960 was a 37,000-ton monster, reflecting the continuing development of South Africa's economy. Regular services were maintained throughout the sixties, although the round-Africa run was dropped in 1961. Then, as elsewhere in the world, passenger figures began to decline and the Union Castle announced that its final passenger sailings would take place in August 1977.

The withdrawal of the Union Castle liners has left the P. & O. as the last of the old-time liner companies still operating regular passenger services rather than providing floating holidays for cruise enthusiasts. Although a considerable amount of P. & O. business is devoted to such cruising, it still sails regularly to Australia and New Zealand, although with decreasing frequency.

The origins of the great P. & O. go back as far as 1836 when two London merchants, Arthur Anderson and Brodie Willcox, chartered their first steamer to trade to Spain as agents for the Peninsula Steam Navigation Company. In the following year, the partners obtained an Admiralty contract to carry mails between Lisbon, Cadiz and Gibraltar. Three years after that, in 1840, the line picked up another contract, this time to take the mails all the way to Alexandria. Here they would go overland to Suez and be carried on to India by ships sailing down the Red Sea. Accordingly the name of the firm was changed to Peninsular and Oriental Steam Navigation Company which was soon shortened to the colloquial 'P. & O.'. By 1854 the line had taken over the Suez to Bombay leg of the route to India, and added services to Hong Kong, Singapore and Sydney.

The opening of the Suez Canal in 1870 produced an upheaval in P. & O. arrangements, and it coincided with the need to re-engine many of its ships with the new compound steam engines, which were more efficient than the simple expansion type fitted in most steamers up until that time. The changes were made, and by the late 19th century, the P. & O. was regarded as 'the Cunard of the East' and almost a pillar of the British state. At this time the line reflected all the military dignity and conservatism of the British officers and civil servants who occupied its first-class quarters on leisurely voyages to and from the Far East. Its ships were never fast or large, but they were well suited to the trade and were always in receipt of Government support in the form of mail or trooping contracts. Before the First World War, few P. & O. ships could make more than 12 knots and even their colors were restrained – black hulls with stone upperworks and black funnels.

P. & O. was never in the van of technical advance during those years, preferring to retain reciprocating engines for all its ships until the late 1920s. However, it did produce twin-screw vessels, not because of speed requirements, but to avoid the embarrassment of a broken shaft in Eastern seas.

But by 1930, P. & O. had gone over to turbine and even turbo-electric machinery and in the thirties produced a series of ships called the 'Strath' class; these were modern express liners.

After the Second World War, P. & O. found itself in a vastly changed world. Britain had left India and the airlines had acquired the mail contracts. But there was much emigrant traffic to be obtained between Europe and Australia, and a series of new liners was built. These were all large vessels and a far cry from the usual P. & O. ship of the early years of the century. Typical of these was the 29,700-ton *Iberia* of 1954.

In 1960 the P. & O. merged with the Orient Line, another longstanding British company on the Far East and Australian routes. The Orient Line could claim beginnings as far back as 1820 and it brought to P. & O. a modern post-war fleet of 28,000-ton ships, such as the *Orcades* of 1948. Like the *Iberia*, these ships were large and fast, as they had to be if they had any chance at all of competing with the airlines. In addition, at the time of the merger, both companies had ordered an even larger steamer for their respective fleets. The Orient newcomer was the popular *Oriana*, 41,900 tons, which produced 30 knots on her trials in October ·1960. *Oriana* has been a successful ship, popular with cruise passengers, and she remains in service in 1977.

The new P. & O. ship was also large,

Flagship of the modern P. & O. fleet, the Canberra *(45,000 tons). Seen here on the slipway at the Belfast yard where she was built in 1961.*

45,250 tons, the largest passenger liner out of the Belfast yards since the *Titanic*. This was the *Canberra* of 1961, produced to a revolutionary design which placed her turbo-electric machinery right aft and gave her a singular appearance. This arrangement allowed spacious accommodation for over 2,000 passengers and the liner is still in commission in 1977. Most of the remaining P. & O. tonnage has been withdrawn. The company still retains a large freighter fleet, however, and it has several modern medium-size cruise ships operating out of American ports.

Such then were the liner companies which, from small beginnings and often in difficult and sometimes dangerous circumstances, built themselves in time into some of the world's most powerful commercial organizations. At the height of their power and fame, the owners of the great liners were the friends of kings and the advisers of governments. Their great ships, created by the skill of thousands of shipwrights and engineers, gave those same craftsmen a stable livelihood and provided employment for further hundreds as crew members or shore staff. The companies created a unique world of the high seas – a world long since vanished – a world in which the greatest luxury and elegance lay but a few decks above the rough and ready quarters in the steerage, which was all that most of the world's travellers were able to afford. In reality, the liners were a microcosm of society in Europe and America during the late 19th and early 20th centuries, and they have faithfully reflected its changes as the 20th century approaches its final decades.

Man's greatest transport machines were nothing if not the floating temples of the 'high life' and as the style and posture of the age of the twenties and thirties departed, so the liners went also. Their legacy is a record of superb technical and commercial achievement, and memories of an age when men of vision possessed the nerve to turn their dreams into reality, and dared to live in a style that matched their dreams and demonstrated their success to all.

CHAPTER THREE

FLOATING CITIES

Liner design and decoration

*Late nineteenth-
century elegance
afloat on the Hapag
liner* Kaiser Wilhelm
II *of 1889.*

Below: the giant
balanced rudder and
two port screws of the
Aquitania tower over
a party of ship-
wrights working on
maintenance duties.
Shipping companies
spent much time
convincing the public
of the giant
dimensions of the
liners. Here we see
Cunard's Queen
Mary in the
improbable setting of
London's Trafalgar
Square.

DURING THE EARLY DECADES of the 20th century, when the age of the liners was at its zenith, any press article or company publicity release on the subject of new ships invariably used the term 'floating city' to describe them. The description stuck in the public mind, because it was apt and mirrored exactly the man-in-the-street's conception of how big a proper superliner ought to be.

The advertising men were abetted in their use of purple prose by the foremost literary figures of the day. 'This monstrous floating Babylon', wrote journalist and social reformer W. T. Stead about the *Titanic*, from whose decks he would in a few days be washed to a freezing death. Theodore Dreiser, writing of his beloved *Mauretania* called her, 'A beautiful thing all told – long cherry-wood panelled walls . . . there were several things about this great ship that were unique.'

For Rudyard Kipling, poet laureate of British imperialism, the big liners were an irresistible subject for his verse at its most buoyant:

'You can start this very evening if you choose,
And take the Western Ocean in the stride
Of thirty thousand horses and some screws!

The boat-express is waiting your command.
You will find the *Mauretania* at the quay,
Till her captain turns the lever 'neath his hand,
And the monstrous nine-decked city goes
to sea.'

So successful were naval architects and interior designers in their pursuit of taking to sea the highest standards of life ashore, that everyone understood stage-star Beatrice Lillie's famous quip as she stepped aboard the *Queen Mary*: 'Say – when does this place get to New York?'

'Vast', 'magnificent', 'fabulous': all these words and more were used to describe the floating palaces of the twenties and thirties. Illustrated magazines published in London, Berlin and Paris carried drawings by technical illustrators all designed to demonstrate the size and excellence of the liners. To achieve the aim of astonishing their viewing public, these draughtsmen sketched the big ships in locations that were familiar to all, but highly unlikely ever to accommodate a 900-foot liner. Thus the *Vaterland* appeared upended on her bow to demonstrate that, if ever she could be maneuvered into such a stance, her rudder would top the highest pinnacle of the Chrysler building, then New York's highest skyscraper.

The *Queen Mary* was set down astride London's Trafalgar Square, with her bow well into Whitehall and her bridge so placed that the Captain could hold a conversation (should he so desire!) with Nelson on his column. One of the more memorable of these fanciful illustrations showed a plan view of the first-class lounge of the *Queen Mary* with not only the three ships of Columbus's entire fleet cosily contained inside, but Sam Cunard's *Britannia* as well! No other piece of liner propaganda demonstrated so clearly the advances in passenger accommodation in the century that had passed since *Britannia*'s first sailing in 1840.

The *Britannia* herself was an age away from the provisions for passengers offered by the sailing packets that were her contemporaries. In the long, bare, between-deck spaces of the packets, all the emigrant received for his money was his passage, a place to wash and drinking water that was quite often not fit for its intended purpose. It usually stank, but this would pass unnoticed in the noisome aroma that pervaded the passenger spaces, which were used quite freely for all the normal processes of the human body.

Modern hygiene techniques had yet to be pioneered, and in this filthy bedlam the steerage passenger would arrive with his food, cooking utensils and his bedding, all of which he was expected to provide for himself. Once at sea, the degrading squalor was compounded by the collapse of all but the stoutest stomachs from chronic seasickness. So bad were conditions in the packets, where passengers

were treated as little better than cargo, that Congress passed two Acts, the first in 1819, and another in 1837, that limited overcrowding and laid down the space allocated by law for each passenger. The British Government followed suit and, in a bid to stamp out typhus, the scourge of passenger shipping, also decreed minimum rations that companies must supply each week to their steerage passengers.

This legislation did at least improve the lot of the emigrant who had, until then, travelled in much the same conditions as the sad victims of the recently outlawed slave trade. Meanwhile, in the well-ventilated (often too much so) cabins on deck, the wealthy enjoyed as comfortable a passage as chintz curtains, horsehair-padded upholstery, teak furniture and the weather would allow.

The crews of these Atlantic packets fared little better (and often worse!) than the steerage passengers. Driven on by brutal discipline and by the knowledge that there were hundreds of unemployed seamen ashore eager to sign on, the topmen of those days lived mainly on hardtack, biscuit and boiled rice. Splendid seamen, in spite of their degrading work conditions, the packet sailors could force a fast passage that might make anything up to $30,000 clear profit per round voyage.

With the coming of steam, there came a new conception of passenger travel. All his life, Sir Samuel Cunard put only the safety of his passengers ahead of their comfort. There were cabins for all the passengers on the *Britannia* and regulations for cleanliness were strictly enforced. Staterooms were to be swept every morning, beginning at 0500 hours (passenger reaction to this crack of dawn activity has not been recorded, but the Victorians were notorious early risers!). Slops were to be emptied while the passengers were at breakfast ('I had a good steak, with a bottle of hock', recorded a satisfied traveller), and bed linen was to be changed every eight days.

All this was a distinct advance, but it did not impress Charles Dickens when he

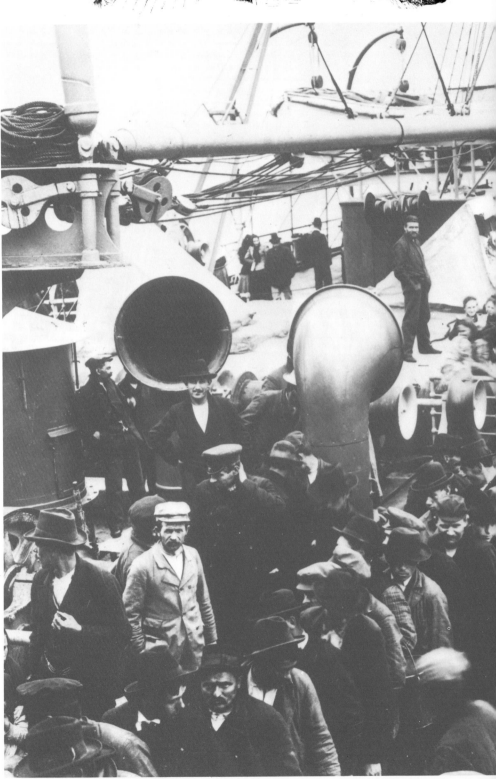

sailed in the *Britannia* for his celebrated tour of America in 1842. The great novelist recorded his experiences on board the Cunarder, using all his undoubted skill as a trained observer, and he has left the best account we possess of a trans-ocean passage in those days. Astute as he was, Dickens fell for the 1842 equivalent of advertising blurb. When booking passage in London for himself, his wife Kate and her maid, Dickens had seen a Cunard handbill showing the saloon of the *Britannia* to be 'a chamber of almost infinite perspective, furnished . . . in a style of more than Eastern splendour'. Such florid description leads one to believe that Dickens was writing with his tongue at least half in his cheek.

Nevertheless, he tells us that the splendid saloon of infinite perspective was in fact 'a long narrow apartment, not unlike a gigantic hearse with windows'. Worse was to follow. Dickens describes sailing day at Liverpool as a 'most extraordinary and bewildering tumult'. Through a throng of struggling stewards and passengers (matters were no better in the 20th century), the Dickens party headed for their cabin, while everywhere passengers swarmed 'down below with their own baggage, and stumbling over other peoples'; disposing themselves comfortably in wrong cabins and creating a most horrible confusion by having to turn out again'.

On arrival at his cabin, Dickens found it to be an 'utterly impractical, thoroughly hopeless and profoundly preposterous box', bunks of which were shelves covered with thin mattresses and a limp quilt. Dickens stormed that 'nothing smaller for sleeping in was ever made except coffins'. Kate's several trunks stood no chance of being accommodated in those narrow confines.

Once the *Britannia* had sailed, Charles and Kate went down with seasickness as the Cunarder's 'every plank and timber creaked . . . like an enormous fire . . . there was nothing for it but bed, so I went to bed'. There he remained for ten days as the *Britannia* fought her way to Halifax

2. THE FIRST DINNER. NOT QUITE SO HUNGRY AS HE THOUGHT HE WAS.

1. BOUND FOR BOMBAY UNDER THE CAPTAIN'S CHARGE, AND OF HIS MANY RESPONSIBILITIES THE HEAVIEST.

4. A PERILOUS JOURNEY.

3. JONES IS NEVER SEASICK, BUT EXPLAINS THAT HE IS SLIGHTLY UPSET BY HIS EARLY BREAKFAST IN TOWN.

6. A LAST GLANCE OF OLD ENGLAND.

5. SMITH FEELS HE MIGHT POSSIBLY SURVIVE ANOTHER TEN MINUTES IF HIS CHARMING COMPANION AND THE ATTENTIVE STEWARDS WOULD LEAVE HIM ALONE.

through gales that crushed her lifeboat, tore planks away from the paddleboxes and left 'the wheels exposed and bare' and the rigging 'all knotted, tangled, wet, and drooping.' A gloomier picture would be hard to look upon. Dickens survived, however, and so did the *Britannia*. Elsewhere, other designers had caught on to the idea that comfort on board sold tickets ashore, and larger vessels with better cabins were already afloat.

Never a man for half measures, I. K. Brunel had designed accommodation into his *Great Britain* that was lavish by the standards of his day. Firmly interposed amidships were the vast engines of the ship, dividing the passenger decks in half. Astern, on two decks was the first class, with dining room below and saloon above. Cabins led off each room at the ship's sides. The forward area was laid out in the same pattern and allocated to the second class. Officers and crew were crowded into the bows on four levels. Amidships in the upper 'tween-decks is probably where the twenty-six water closets (referred to by Captain Claxton) were located, although no record survives of their exact position. But water closets were a distinct improvement on chamber pots! On the

other hand, the cabins remained on the small side, being exactly six feet square.

The *Great Britain* made use of large quantities of carpet and the newspapers reported that over 1,000 yards of best-quality Brussels (but made in Bristol) had been laid in her passenger spaces. Claxton, as managing director, quickly pointed out that the company did not consider this to be needless expense simply to procure 'gaudy decoration'. Mirrors were used throughout the ship to increase the impression of space, and there was a combined boudoir and sitting room, reserved exclusively for the ladies. Of the first-class dining saloon Claxton wrote, 'This is really a beautiful room. Its fittings are alike chaste and elegant. Down the centre are twelve principal columns of white and gold.'

For all this comfort, Brunel's company proposed to charge 35 guineas for the best first-class cabins on a single voyage, ranging down to 20 guineas in the second class. Cunard charged at that time a flat 30 guineas anywhere on his ships.

Where Brunel and Cunard led, others were quick to follow and improve standards, not because they saw themselves as social reformers of life at sea, but

because one well-publicized innovation could attract passenger patronage away from the ships of a rival line. Of true public reform, little impact would be made on the rigid first, second and steerage classes and their absolute segregation until well after the First World War. Afloat, even more than ashore, you got exactly what you could pay for!

The history of internal décor aboard ship in the 19th century was made partly by inter-company rivalry and partly by the adaptation of new inventions to maritime conditions.

No sooner was the Cunard fleet established on the Atlantic than the first challenge to its hegemony came from the American Collins Line. Collins set out to win the trade back to American ships and he judged that luxury aboard was part of the answer to his problem. Fresh food, he judged, was another. Shipboard fare in the mid-19th century was plain to a degree, and contained little that was perishable after four or five days out from land. Fresh milk was to be had, as ships like the *Britannia* carried at least one cow, but it was not plentiful for obvious reasons. So Collins planned an ice room for his *Atlantic* with capacity for 40 tons

*Charles Dickens'
cabin on the
Britannia, which he
described as 'an
utterly impractical,
thoroughly hopeless
and profoundly
preposterous box'.*

of frozen water that would take a fort-night to thaw out and preserve all kinds of previously unavailable delicacies for his clients. The *Atlantic* was able to offer fish dishes among the nine courses and more than forty items were available on her usual dinner menu. If you travelled with Collins, you ate in style. 'No vegetable, fruit, game or other rarity that can be kept for fifteen days in large masses of ice, is neglected', wrote one satisfied American passenger.

Collins' ships were steam-heated and possessed bathrooms, an unheard of amenity until then. Another innovation which showed E. K. Collins' flair for style and taste was a fully fitted barber's shop, so that gentlemen might appear well groomed as if they were still striding along Piccadilly or Broadway. The public rooms displayed even more brocade and the plush upholstery so beloved of the age. Marble-topped tables were every-where, as were the now ubiquitous mirrors. The cabins, however, were no larger than on the Cunard ships and some passengers complained of conditions so cramped that dressing was something of a problem. Men had to 'jump from the bed shelf to get into their pantaloons'. Another drawback of the Collins vessels was the continuous vibration caused by the incessant engine beat as the American skippers strove to make ever-faster passage times.

Tiny cabins and vibration apart, Collins' bid for the Cunard traffic succeeded and, in two years from the maiden voyage of the *Atlantic*, his ships were carrying twice the number of passengers that remained loyal to Cunard. There was only one thing wrong. Collins' high standards of comfort and culinary excellence cost far more to provide than was recoverable from ticket money and eventually, after several accidents to his ships had dented public confidence, Collins lost his government subsidy and his line folded. But before he went, he had created the early beginnings of the transatlantic liner tradition of luxury, style and epicurean delight that was to grow more grandiose

and more attractive of public attention.

If Brunel had made the *Great Britain* an attractive lure for passengers, the British inventor really let himself go in his behemoth *Great Eastern* of 1859. Brunel was a man who eschewed the verbose language of his time when the Victorians never used one word where five would do. Offered a list of names considered suitable for such a ship on the morning of her proposed launch day, Brunel testily brushed the proposals aside with the riposte: 'You can call her Tom Thumb as far as I'm concerned!' Brunel was a demonstrator and innovator if nothing else, and on the *Great Eastern* he set a pattern that would still be followed a century later.

The *Great Eastern* had a capacity for 4,000 passengers and a range designed to take her to Australia with just one coaling stop at Calcutta. Brunel made her smallest cabins almost double the size of the largest existing on any other ship of the day, and each one contained a wash basin, rocking chair, dressing table and a sofa that screened a semi-hidden bath. This last luxury was supplied with hot and cold running water. Only the integral WC, air conditioning and electric lighting were missing; otherwise, the *Great Eastern* would have possessed the total 'private facilities' today demanded by the cruise passenger of the 1970s.

Outside the cabins, the public rooms of the *Great Eastern* were a riot of red velvet upholstery, gilded columns and domestic statuary whose nameless creators reflected the taste and spirit of the age by their choice of chaste and pious subjects for their art.

Despite all this, the *Great Eastern* had her drawbacks. Apparently her hull vibrated like a monstrous violin and a piano played in the saloon could create hell for some luckless passengers half the ship's length away. The other problem was that she rolled heavily and in a gale she went all over the sea, causing havoc between decks. Furniture, unsecured to the deckhead, would plunge the length of the saloon, crash into the side mirrors and

shower glass splinters on any unfortunate who happened not to have sought the relative safety of his cabin.

Where the *Great Eastern* led, others followed. Limited electrical equipment had been fitted during the late 1860s in the White Star *Britannic* where electric bells summoned stewards for the first time. In Cunard's *Umbria* and *Etruria* of 1886, oil lamps were abolished at last and the brilliance of electric lighting shone out, derived from four generators.

By the eighties, a subtle change had occurred in the ocean passenger. The emigrant was as numerous as before and continued his great exodus westward. But now the new industrial plutocracy of the eastern states of the United States began to flow in the opposite direction in increasing numbers. The dangers of an Atlantic crossing receded as the century aged and ships grew in size and power, and more people who could afford the passage travelled each year to seek the pleasures and culture of the old world. By this time, the liners offered all the comforts of a first-class hotel, but the new clients looked for more, seeking the same standards that were provided daily in their homes along Fifth Avenue. They

CUNARD LINE
"LUSITANIA" "MAURETANIA"

FIRST CLASS LOUNGE

lived in an ambiance which took plush surroundings for granted, where *haute cuisine* was expected and received with critical appreciation, and where anything less than opulence was considered second-rate. It was this group that ensured that the next generation of liners would graduate from seagoing hotels to the 'floating palaces' of the 20th century.

The transformation began in Cunard's *Campania* of 1893 where the decorators achieved a style in mahogany, sandalwood and stained glass that could only be described as English Baronial, and such a style needed bedrooms and parlors straight out of an English stately home. Cunard, claiming that this was 'a silent sermon in good taste', were only too happy to oblige, and the resulting effects on tradition-starved Americans was impressive indeed. They flocked to join *Campania*'s and her sister *Lucania*'s ever full passenger lists. From the baronial quarters of the *Campania* to the palatial *Mauretania* and *Lusitania* was now but a short step for Cunard.

In just thirteen years, the size of the Atlantic liner trebled, and the consequent increase in space for the passenger allowed a new grandeur afloat. The new century brought new tastes, and the décor on the *Mauretania* had lost much of the heavy Victorian overtones. The public rooms on the new liner, set around the uptakes of four enormous funnels, lost nothing in the process. Rather, the restrained panelling of polished brown mahogany in the first-class lounge gave a reassuring atmosphere of stability, and everywhere the panels blended into carvings skilfully executed by 300 Arab craftsmen specially brought over from Palestine.

By and large, the Cunard directors, like most other British shipping magnates, planned their liners as they did their own homes, and both the *Mauretania* and the later *Aquitania* reminded the British aristocracy of stately premises in the English shires. These homes contained all manner of artistic styles and the ships were much the same. The library of the

The main first-class stairway on the Olympic. *The stair covering has yet to be laid but the elegant panelling is complete, and the carving symbolizes Honor and Glory crowning Time. A similar theme was used on the* Titanic.

Mauretania was based on Louis XVI and the bookcases were copied from originals in the Trianon. The main lounge was also eighteenth-century French, but the first-class dining room capriciously jumped back two centuries to the time of François I. The Verandah Café, a new feature offering a de luxe *à la carte* menu (not included in the fare!) was stolidly English, derived from the Orangerie at Hampton Court.

The world's press was called in to preview all this magnificence and pronounced it fit for the most demanding sybarite. Almost as an afterthought was there a mention of the elevators, two of which were fitted in the main staircase well. They were the first in any large British ship, although the German *Amerika* had introduced the first set afloat a year earlier.

So luxury went to sea in the *Mauretania*, and immediately the air of elegant repose took a sharp knock when the ship made her first high speed run. The truth had to come out: *Mauretania* rattled vigorously at anything like top speed. Although the fitting of new four-bladed screws in 1909 (in place of the original triple-bladed variety) cut the vibration to acceptable limits, the rattle was there for all the *Mauretania*'s subsequent career.

The first decade of the 20th century saw the beginnings of the penultimate in superliner design. Following the Cunard initiative with the *Mauretania*, the White Star and Hamburg-Amerika reply was inevitable. The result was six big ships, four of which would be pre-eminent on the Atlantic for the next quarter century.

In the fall of 1907, as the *Mauretania* and her sister were making their startling maiden voyages and receiving acclaim as 'the wonders of the age', two men met over the dinner table of a house in London's fashionable Belgrave Square. One was Bruce Ismay, chairman of the White Star Line and a major figure in Morgan's shipping empire. The other was Lord Pirrie, shipbuilder extraordinary and boss of Harland and Wolff, and responsible for the creation of more great

passenger liners than any other constructor in modern maritime history. The results of the late-night discussions of this powerful duo were the great White Star monsters *Olympic*, *Titanic* and *Britannic*, half as large again as the two Cunarders and the greatest ships yet. Subsequent events would also combine to make the trio the most tragic class of any liners, slaying thousands and heralding the demise of the White Star Line itself.

All that was in the future when the *Olympic* sailed on her maiden voyage from Southampton on 31 May 1911. The White Star had moved their prestige passenger terminal to the Hampshire port four years previously. London was just one hour and a half away by boat train, and the great French port of Cherbourg lay eighty miles of calm sea away across the southern horizon with its access to continental Europe. After the World War, Cunard's prestige ships left Liverpool, and Southampton became the home of all British superliners. As the *Olympic*'s first passengers looked about them, those fortunate to be in first class were able to gaze on styles that included not only the now obligatory Louis-Seize but also Quatorze and Quinze as well, backed up for good measure with Empire, Italian Renaissance, English Jacobean, Queen Anne and Georgian, and even a mock Adam fireplace or two.

Those passengers, and they were many, who could discern the varied styles of the ship's appointments, commented favorably on the elegance around them. Those who could not were no doubt impressed that the Louis-Seize dining room was the largest then afloat and that fact more than made up for the floor covering of dull red but practical linoleum. Carpeting for diners would arrive with the *Titanic*. The White Star made less use of panelling in the public rooms than Cunard and so *Olympic* had a much 'lighter' air about her than the Victorian atmosphere that pervaded the earlier ship.

For the energetic, the *Olympic* was a definite advance on any previous offering.

*The first-class
dining room on the
Olympic, carried out
in Louis-Seize style,
complete with gilt-
banded columns and
chintz-backed chairs.
The same style was
used on the Titanic,
but her dining saloon
was carpeted through-
out, unlike the
Olympic, which in
her early years 'made
do' with linoleum.*

THE DECOR
OF THE
GREAT LINERS

The delicacy of design in the
Normandie, certainly the greatest
French liner of all time, offset the
sheer immensity of the ship. The
leading figures of French interior
design were commissioned to render
the liner a floating showcase of the
best in the decorative arts of the
thirties. In contrast to the flair and
elegance of the interiors of the
Normandie, those of the British
Queens seemed to express a sobriety
which was perhaps more in keeping
with the national outlook during
those years.

*Above : bas-relief in sprayed nickel in the
tourist lounge of the* Queen Mary.

*Below : fireplace ornament in the smoking
room of the* Queen Elizabeth.

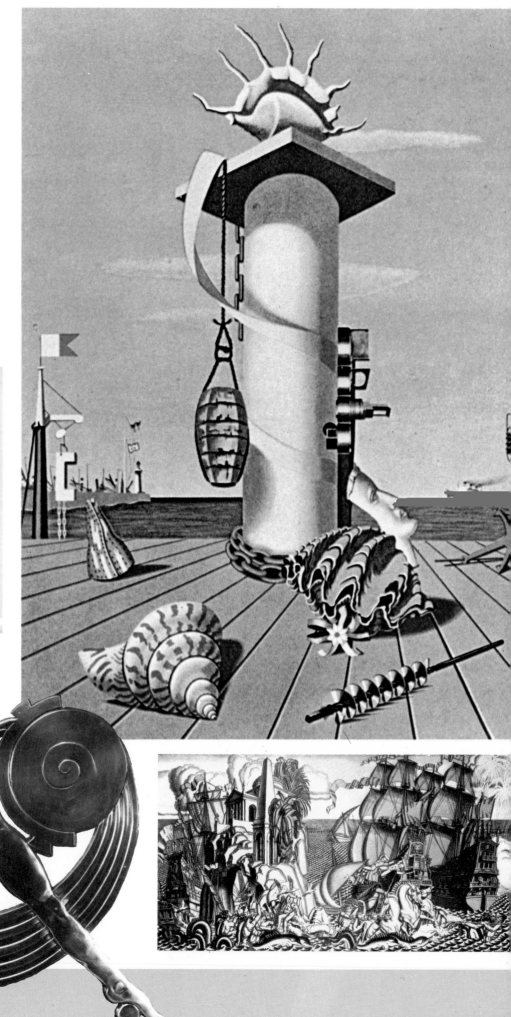

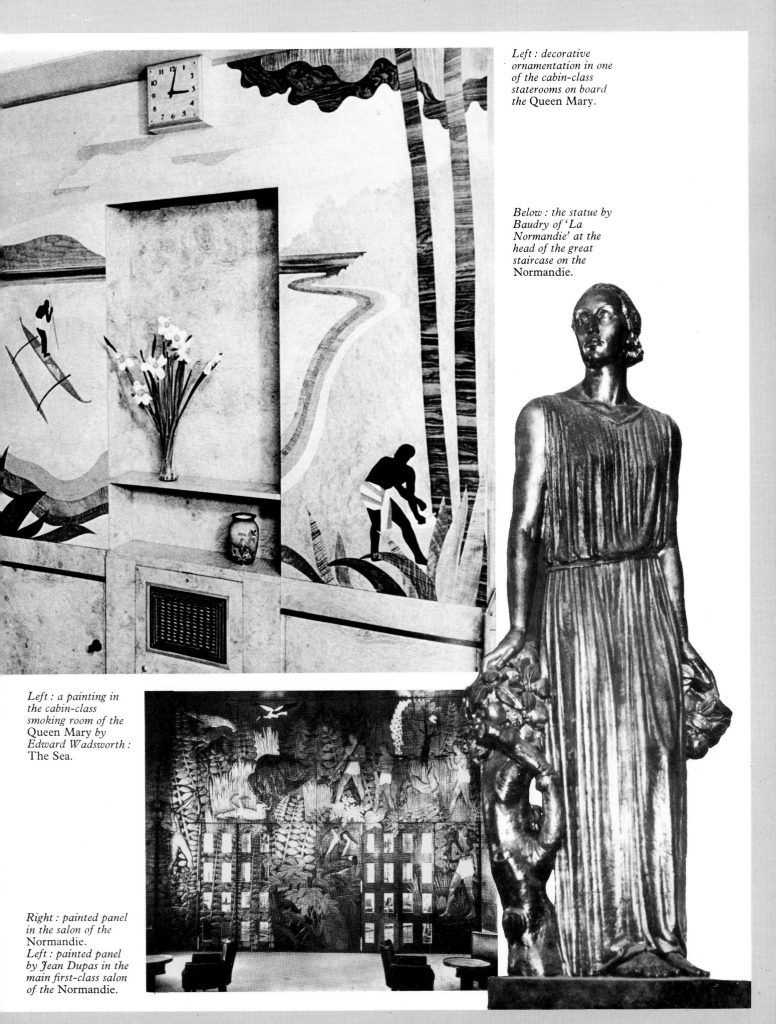

Left : decorative
ornamentation in one
of the cabin-class
staterooms on board
the Queen Mary.

Below : the statue by
Baudry of 'La
Normandie' at the
head of the great
staircase on the
Normandie.

Left : a painting in
the cabin-class
smoking room of the
Queen Mary by
Edward Wadsworth :
The Sea.

Right : painted panel
in the salon of the
Normandie.
Left : painted panel
by Jean Dupas in the
main first-class salon
of the Normandie.

On the boat deck, a fully equipped gymnasium, complete with resident instructor, awaited those who felt the necessity of joining the emergent keep-fit mania of Edwardian times. Electric horses and bicycles, rowing machines and vibrators were available to help trim off the excesses of the restaurant and allow a fit and alert commuter to land at New York, ready for all that the business world of that city could hurl at him. Down below there was a full-size squash court and, marvel of marvels, the first ocean-going swimming pool, an attraction without which no ocean liner would ever again be built. To make perfection perfect, alongside the pool was another seagoing first, a Turkish bath and sauna, complete with resident masseuse. 'The *Olympic* is a marvel,' pouted Ismay, not without justification. His and Lord Pirrie's grand design had been more than realized with this great ship which, due to the skill of Pirrie's naval architects at Belfast, contained all her luxurious grandeur in a slim hull of 45,000 tons that rode steadily on the highest seas, and at the same time gave the four-funnelled giant an appearance that has seldom been rivalled.

Due to the decision of the White Star Line to concentrate on passenger comfort as opposed to record passages, the *Olympic* was not the rattler that the *Mauretania* turned out to be. Neither were the three big German liners which, coincidentally with the White Star trio, Albert Ballin was creating for the Hamburg-Amerika Line. Germany was renewing her challenge on the Atlantic with ships that would again break records, but this time for size rather than speed. The

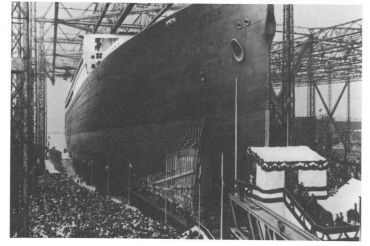

Left: Kaiser Wilhelm II about to launch the Imperator, *23 May 1912. Above: the Ritz-Carlton Grill on the* Imperator. *The overall style is eighteenth-century French, but Mewès added touches of various other periods where he thought necessary. Right: Hapag publicity pictures of passengers relaxing in front of one of Mewès' mock Adam fireplaces.*

Bottom left: the bronze Imperial Eagle figurehead carried by the Imperator *on her early voyages. The giant bird sustained gale damage off Cherbourg and was removed on the liner's first refit.*

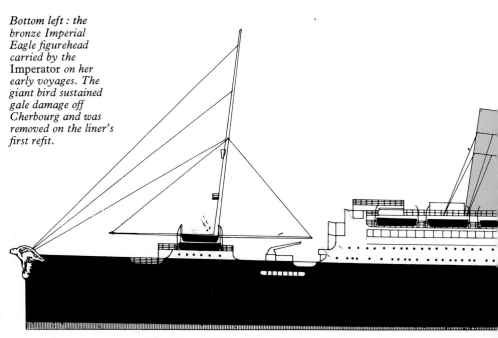

ultimate in size, opulence and comfort in liners was the objective of Ballin and his collaborators.

Once again, they had the full and enthusiastic backing of their Emperor. The interest shown by Kaiser Wilhelm was such that the Hapag board made a last-minute switch of names for the first ship. She was to have emerged as the *Europa*, but in due deference to her Imperial patron who proposed to launch the ship himself, she became *Imperator* and was followed by *Vaterland* and *Bismarck*. The names were an inspiration for those with a vision of German expansion, but they had an ominous ring for the rest of Europe, who had come to distrust the Kaiser's international ambitions. As if to deliberately confirm all the suspicions of those who kept a wary eye on German imperial pretensions, Hapag fitted up the *Imperator* with the most amazing decorative device that ever went to sea. It took the form of a monstrous bronze-gilt Imperial eagle, 30 feet in length, which grasped in its claws a globe mounted with the legend 'Mein Feld ist die Welt'. Germany's critics felt they knew exactly what that meant, but the

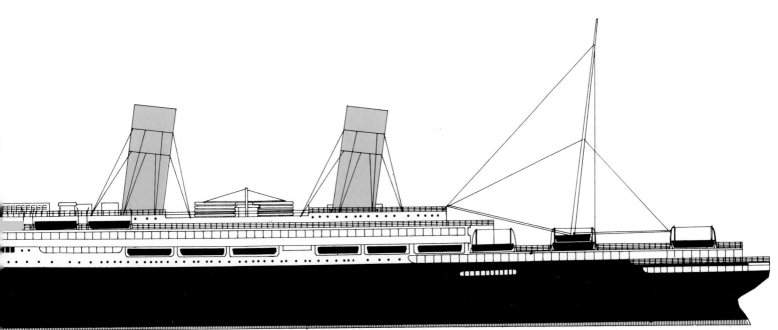

The Gay Twenties on board Statendam III.

reason for the great figurehead had no more sinister explanation than that the Germans needed the extra length to maintain the *Imperator*'s superiority in size over Cunard's new *Aquitania*, then being completed at Clydebank. No matter what the motive for the bird's appearance, no one could deny that it had a most lugubrious look in its large dead eyes and that the total effect was most depressing on a ship which was already being criticized for her allegedly heavy appearance.

Once on board the *Imperator*, however, any discordant notes about her external appearance turned into a chorus of praise for the magnificence of her internal accommodation and décor. If the *Mauretania* and *Olympic* had taken the stately homes of England to sea, the *Imperator* had exceeded all that by providing an imperial palace afloat. To achieve this, Ballin had gone outside the world of ship construction to the internationally-known, interior-decoration consultant Charles Mewès, whose main design offices were based in Cologne.

Mewès had come to public notice because of his collaboration with the great hotelier César Ritz, whose name has long passed into several languages as a synonym for opulent splendor. The pair had already produced the great hotels, including those in Paris and London, which were to make the Ritz chain the most famous in the world. Then Ballin invited Mewès to take on the designs for his liner *Amerika* of 1903. Mewès' work was so successful that the *Amerika* became the smartest ship of the Atlantic until the *Mauretania* came out in 1907. It is said that Ballin reached his decision after he had been one of the first diners in the splendor of London's Ritz Carlton Grill. The food made a great impact on Ballin's palate.

On the *Amerika* he planned the first *à la carte* restaurant, separate from the ship's dining saloon, that would serve meals at any hour for an extra charge. This concept of an exclusive club for an already patrician clientèle was a shrewd assess-

ment of human nature that proved profitable. César Ritz landed the contract for providing the bill of fare, cooking and serving it. To do so, Ritz sought the aid and expertise of no less a grandee of *haute cuisine* than the great Auguste Escoffier himself. The only instruction that Ritz gave to his maître d'hotel was that the food should be exactly as served in the Ritz Carlton Grill in faraway Paris. The Ritz staff obliged and in the years that followed, several of the world's great *restaurateurs* operated concessions afloat. The concession on the *Titanic* was operated by the London firm of Gattis, and Luigi Gatti and all his staff were drowned when the liner foundered.

The success of the *Amerika* made Mewès the inevitable choice as consultant for the décor of the *Imperator*. He was now at the height of his powers and,

together with his equally brilliant young partner Arthur Davis, he created new standards in the *Imperator* that left their influence on all that followed after. Mewès was an Alsatian and therefore nominally a German citizen, as his native province was then a temporary part of the German empire. But he retained a Gallic flare for design which made the grandeur of his concepts outstanding, while remaining places in which people could live. The first-class rooms on the *Imperator* became the Ritz at sea and much more besides. The absence of a fourth funnel allowed more space for the grand lounges that flowed from Mewès' drawing board. To the overall eighteenth-century French style, he added Adam touches that involved the use of quantities of marble.

The effect was stunning. Below also, using his designs for London's Royal

Automobile Club building in Pall Mall as a prototype, Mewès created a swimming pool that was quite literally a marble hall and he repeated the design in the two ships that followed. He also got his way in a long-standing dispute with Ballin's naval architects. Mewès had long-argued that, to obtain the best layout of a ship's public rooms, the huge funnel uptakes that carried away fumes from the boilers should be divided and routed up the side of the vessel, thus giving long central vistas down the length of the decks. In the *Vaterland* and the *Bismarck* he got his way and a great sense of added spaciousness was thus obtained. The use of this uptake layout was repeated in the French Line's *Normandie* with particularly spectacular results.

The *Imperator* was ready for service in May 1913 and on trials she scared her

The Vaterland leaving her builder's yard, Blohm und Voss at Hamburg. Her career was interrupted after only four round trips to America by the First World War, and when the war ended, she became the United States Lines' Leviathan.

The swimming pool on board the Vaterland, complete with marble columns reaching through two decks. The prototype for this design may be seen in London's Royal Automobile Club.

QUEEN MARY
A synonym for the greatest

Known mainly as a 'happy' ship, the *Queen Mary* perhaps lacked something of the dash and flair of the great French liners of the same age. Nevertheless, she commanded a special place in the heart of the English nation at a time when national pride and prestige were flagging. Her sheer hugeness became the synonym for everything grand and grandiose, and interest in her every aspect was intense, as we can see from this display of cigarette cards of aspects of her sober interiors.

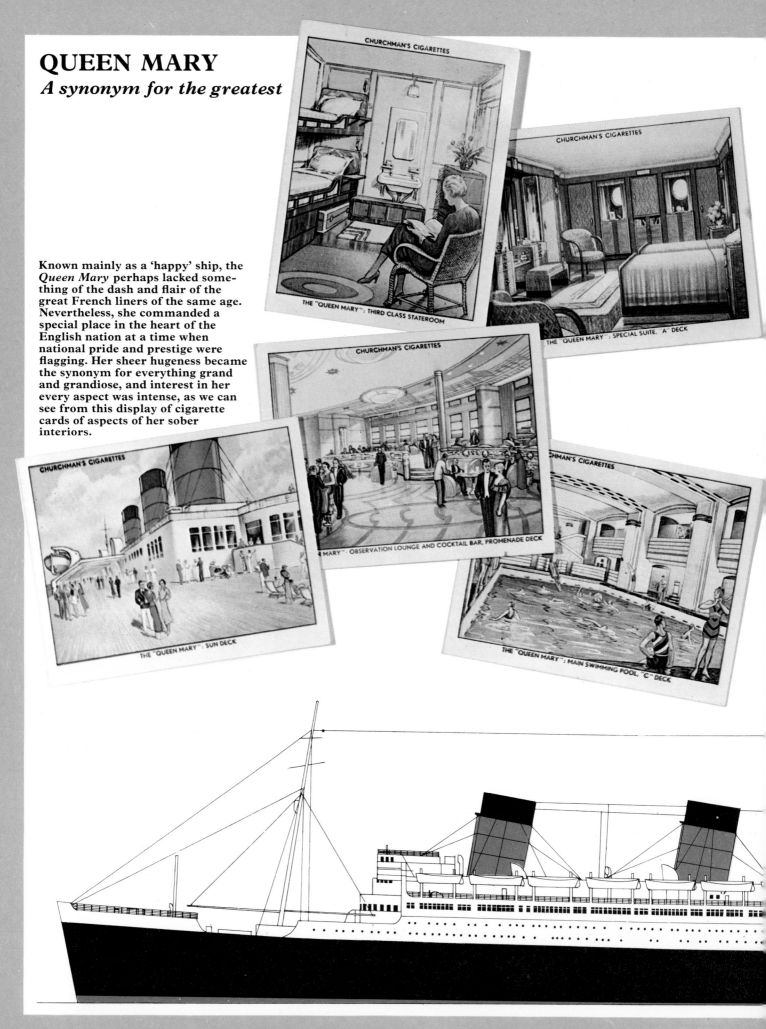

THE "QUEEN MARY": THIRD CLASS STATEROOM

THE "QUEEN MARY": SPECIAL SUITE. "A" DECK

MARY": OBSERVATION LOUNGE AND COCKTAIL BAR, PROMENADE DECK

THE "QUEEN MARY": SUN DECK

THE "QUEEN MARY": MAIN SWIMMING POOL. "C" DECK

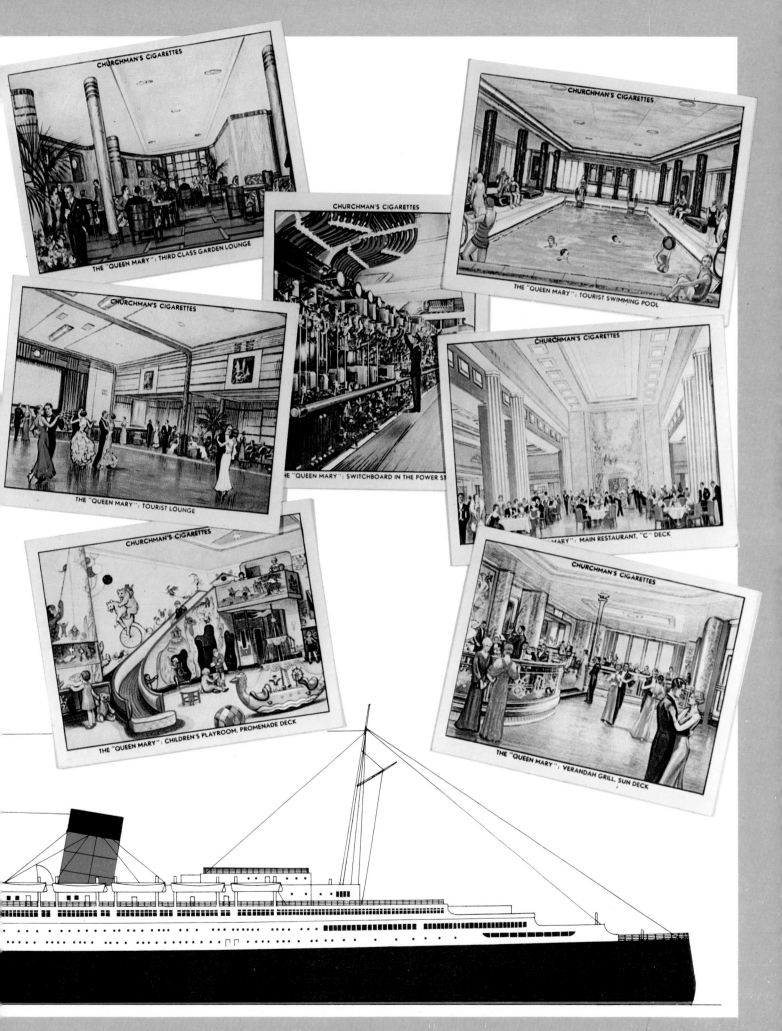

CHURCHMAN'S CIGARETTES

THE "QUEEN MARY": THIRD CLASS GARDEN LOUNGE

CHURCHMAN'S CIGARETTES

THE "QUEEN MARY": TOURIST SWIMMING POOL

CHURCHMAN'S CIGARETTES

THE "QUEEN MARY": SWITCHBOARD IN THE POWER ST...

CHURCHMAN'S CIGARETTES

THE "QUEEN MARY": TOURIST LOUNGE

CHURCHMAN'S CIGARETTES

...ARY": MAIN RESTAURANT, "C" DECK

CHURCHMAN'S CIGARETTES

THE "QUEEN MARY": CHILDREN'S PLAYROOM, PROMENADE DECK

CHURCHMAN'S CIGARETTES

THE "QUEEN MARY": VERANDAH GRILL, SUN DECK

111

*Edwardian elegance
on board Cunard's*
Aquitania *of 1913.
Top left: part of a
private suite. Top
right: a corner of the
first-class smoking
room. Center left:
the ship's pool.
Center right: the
ballroom. Bottom
left: the Palladian
lounge, claimed to be
the most elegant room
afloat. Bottom right:
the first-class
stairway.*

*The third-class
dining saloon on the*
Vaterland*; it is now
1914 and there are
even stewards in the
steerage.*

crew stiff by her tendency to roll dra-matically in anything of a seaway. The stark fact was that the *Imperator* was top heavy. At the end of her first season, hundreds of tons of top hamper were removed, her slim funnels were cut down and ballast added below. The grotesque figurehead survived this stabilizing pro-cess but part of it was torn away by heavy seas off Cherbourg and the re-mainder decently removed when the ship returned to Cuxhaven.

While striving for perfection in the service he provided for his first-class passengers, Ballin never forgot that his line's prosperity rested equally on the emigrant traffic that filled the steerage quarters on his ships. The *Imperator* car-ried 1,750 steerage passengers alone and Hapag had greatly improved their lot. By this time, there were smoking rooms and dayrooms, and the food provided was plentiful, if not exactly famous for its variety. Also in these post-*Titanic* days, every soul on board was guaranteed a

place in a lifeboat. The old-style emi-grant dormitories of forty or more berths were now a thing of the past. On Cunard's *Aquitania*, which came out a year after *Imperator*, most of the 1,998 third-class passengers were put up in two- and four-bunk cabins. The class barriers were still as rigid as ever and another fifty years would pass before one-class ships became a permanent feature of oceanic travel. Needless to say, *Aquitania's* interior in the first class was equal to anything the age possessed and her Palladian Lounge by Mewès' partner Davis represents the apotheosis of Edwardian shipborne décor.

Of all the causes of social and artistic progress, it has to be admitted that war is one of the most effective. Thus it was, in the short years when the big liners were caught up in the First World War, the upheavals created by the conflict so affected public taste that, when the ships returned to service, they had taken on a dated appearance, looking like Victorian museums with heavy Teutonic overtones.

To a world weary of war, and seeking a spiritual release from the constraints imposed by nineteenth-century society, the simplicity of line and lavish use of color, albeit in restrained pastels, opened up exciting new concepts in design.

Like many another vibrant and violent decade, the twenties had many features that are best forgotten, but in 1927 there appeared from the Penhoet yard at Saint Nazaire a ship that deserves to be remem-bered: the *Ile de France*. Heading the French Line at the time of the planning stages of the *Ile* was Jean Piaz who decreed, 'To live is not to copy, it is to create'. With this injunction ringing in their ears, the technical staff of the French Line called in thirty of Europe's top designers. The results reflected the age. Surprisingly harmonious, they did not impress the older generation of Atlantic travellers, but the young, or anyone who had the least aspiration to the status of avant-garde, flocked to the new ship. They relaxed in lounges decorated by Jeanniot, Bouchard and Saupique, walked round statues by Baudry and dined in a 700-seat marvel of a dining room by Pierre Patout, who used Pyrenean marble in three shades of grey.

Elsewhere, the *Ile* contributed her measure to the list of 'firsts' by sailing with the first ocean-going, consecrated Roman Catholic chapel. More than any other ship, the *Ile* endeared herself to thousands of passengers, not for her size or speed, or even her comfort, but the sheer French élan that permeated shipboard activity. Small wonder that the *Ile* went into several popular songs of the thirties, including Coward's 'These Foolish Things'.

Contemporary with the *Ile de France* were the two German Blue Riband challengers *Bremen* and *Europa*. Com-pared to the new Germans, with their bulbous bows, sleek modern hulls and squat funnels, the *Ile* looked positively ancient, but she scored over them in respect of interior attraction. In their ships, the German designers had taken the new fashion for economy of line to

ILE DE FRANCE

CHAMBRE A COUCHER
Suite 467

Top: the Grand Salon of the Ile de France.
The lounges on the Ile *were by Jeanniot,
Bouchard and Saupique, and were a
triumphant breakaway from all that had gone
before.*
Above: first-class bedroom on the Ile de France.

excessive lengths and there were many
who found much of the *Bremen*'s décor
approaching the clinical. Notwithstand-
ing this rather bare appearance, the
Bremen introduced the age of new
materials in plastic and synthetic materials, and started a trend that would lead
to the eventual fireproof ship.

The year 1930 is a turning point in the
history of the floating palaces. In that
year, six big liners were being planned
in the drawing offices of Europe, two each
for Britain, France and Italy. It was then
that British designers appeared to lose
their way and indulged in rectilinear
forms in chocolate brown, enclosing yard
after yard of walnut panelling. It first
appeared in Canadian Pacific's *Empress
of Britain* of 1930. Frank Brangwyn and
Sir John Lavery designed rooms that

were magnificent in their scope and
breadth, but which had more than a touch
of the popular London hotels of the day.
Much of the same approach went into
Cunard's *Queen Mary*. As a ship, designed
for use on the Western Ocean in all
weathers, the *Queen Mary* had no equal,
unless it was her equally seaworthy sister,
the *Queen Elizabeth*. Their interior design
was less successful and, although both
were popular ships and Cunard service
the best anywhere, the surroundings on
board were just too 'super cinema' for
any period except the thirties.

The Italians took no risks on the *Rex*
and the *Conte di Savoia* and went back to
the grand age of Italian Baroque. The
ceiling of the main lounge of the *Conte*
would not have disgraced the Sistine
Chapel. Designed for the southern route,

114

Above : Edmund Dulac's startling Chinese lounge on Canadian Pacific's Empress of Britain *of 1930.*
Far left : the Bremen's *stylish swimming pool.*
Left : the strong rectangular lines of the Bremen's *grand main lounge.*

115

with its sunshine, the Italians sported open air lidos and sun terraces.

It was from France, however, that in 1935 came a liner with the most valid claim of all to perpetual greatness: the fabulous *Normandie*.

Designed by an émigré Russian, Vladimir Yourkevitch, whose career began in the Petrograd Admiralty, the hull of the 83,000-ton giant introduced features which are now commonplace in naval architecture. A bulbous forefoot, slimmed almost to razor edge at the waterline, then soared into a majestic prow that flared away into the streamlined sides of a storm bow. The effect was instantaneously dramatic and, capped by three huge cantilever funnels, gave an appearance of powerful majesty. The unsupported funnels, with the ventilators concentrated in their base, allowed open deck space that would not be seen again for thirty years. These same funnels had another feature already mentioned: their uptakes were divided and so allowed a sweeping procession of public rooms over 500 feet of the liner's centre section.

Again, the French Line summoned France's best artists to the task of creating an expression of all the best in contemporary national style and they succeeded in a manner that occurs but once in an age. The design concepts in the *Normandie* were immense by any standards. Yet a delicacy of line offsets sheer size as one studies the illustrations that are all that now remain to remind one of this great ship from the thirties. The great lounge and equally spacious smoking room, the work of Acon and Patout, could be merged into one enormous salon. Forward of this was a fully equipped theatre, and astern a central staircase swept up to another hall and on to the veranda grill overlooking the stern terraces. On the deck below, the dining saloon was a *tour de force* of tinted glass which was used in great quantities as wall covering and also for internally lit pillars which ranged almost the length of this marvellous room. It turned out to be slightly longer than the famed Hall of

Mirrors at Versailles. Lower still, the swimming pool was a lovely contrast of brass fittings and blue marble, and the chapel, the largest ever fitted in a liner, was in Byzantine style, with magnificent marble side panels.

Much use was made in the *Normandie* of carved bronze doors and the dining saloon possessed no less than eight sets. The huge main doors were gilt over bronze and led out to a stairway that swept up to an entrance vestibule lined with Algerian onyx. One can imagine the style and flair with which lady passengers must have swept down those stairs to pause fleetingly at the entrance, giving the whole room the opportunity to admire their finery.

The career of the *Normandie* lasted but four years as she did not survive the Second World War. When the time came for the French nation to produce another superliner, it was a ship with the same Gallic elegance in her profile. The *France* of 1961 had two winged funnels that quickly became her trade mark, but Atlantic buffs confessed themselves disappointed with the public rooms on this successor to the *Normandie*. Her dining room, however, was a splendidly muted study in gold and a vast amount of carpet fitted throughout gave her an air of luxury that the *United States* and the *Queens* lacked.

The *United States* was the first large liner to appear after the Second World War. By then the great age of ship décor was over and the time of strictly functional appointments had arrived. Aboard the *United States* everything had a clean efficient look, similar to that in any of the functional hotels operated by American chains around the world. The use of light alloy was pursued even into ornamental sculpture that was attached at random to the plain green walls of the ship's many corridors. After the initial interest in these mainly patriotic motifs had waned, the impact became one of boredom.

In fact, so much did hotel practice dominate interior decoration in the sixties that the official brochure issued to des-

cribe the Italian liner *Michelangelo* actually drew attention to the similarity of her three foyers to those that could be found ashore. Much of her décor, executed in modern flameproof materials, nearly all of them synthetic, reminded the observer of modern Italian resorts.

When the *QE2* was approaching completion in 1968, Cunard's publicity releases did nothing at all to reassure the discriminating passenger. 'Ships have been boring long enough!' shrieked the billboards and the curious waited with something like trepidation to see just how Cunard were about to liven up a way of life for which, after all, they were as responsible as anyone. In the event, the worry proved unnecessary. Given that the 20th century was well into its second half and that standards were rapidly changing worldwide, the putative last of the superliners turned out – on the whole – to be a triumph of good taste. As Cunard have freely admitted, the ship was planned by James Gardner's design team to be a floating resort and had therefore to accept her share of resort facilities in the shape of discos, clubs and coffee shops. But the bold shapes and colors that have gone into her public rooms make her the finest ship afloat today. Her Double Room, dominated by red carpeting and plum suede walls, by Australian Jon Bannenberg, and Michael Inchbald's striking Queen's Room will compare with anything that has gone before. The Columbia Restaurant, stretching full across the ship, has a spacious air aided by the large picture windows on either side, but it still retains a raised platform at the entrance.

The *QE2* sails on into the seventies, bearing little resemblance to many of the great nautical interiors that have preceded her. Yet the same values of decorative excellence and epicurean delight sought by Ballin, Mewès and Escoffier, and brought to the high peak of refinement in the *Normandie*, still sail with her and, let us hope, will continue to be available for all those who care to seek them out as the years go on.

CHAPTER FOUR

THE HOSTILE ELEMENTS

Great liner disasters

*Burning furiously, the
French liner
L'Atlantique, adrift
and out of control in
the English Channel
on 4 January 1933.*

THE GREAT OCEANS are the world's loneliest places. More than mountains and deserts, the sea is the most hostile element that man has to encounter on his home planet. The mountaineer who assaults the south face of Everest by the direct route, or the camel driver who crosses the Gobi are at least on terra firma and stand an even chance of rescue if things go badly wrong. The ocean provides no such support, however frail, and no man can survive alone and unaided for more than a few hours. The great monotonous wastes of turbulent water make a small lifeboat almost impossible to find and it is small wonder that, until the Middle Ages at least, European man found little encouragement to sail the Western Ocean.

When at last he had disabused himself of the idea that anyone sailing to the far west went over the edge of the world, the early navigator still had to contend with gales, fog, shipwreck, fire, starvation, icebergs, scurvy, mutinous crews, and the hostile indigenous native, should he be so fortunate as to survive all the other dangers and arrive at his intended journey's end. As the centuries progressed, ships became larger and safer, but even in the early 19th century, insurance companies worked on the basis that one ship out of every six sailing from European ports would never reach her destination. Many of the ships were the infamous 'coffins', loaded to the gunwale with emigrant families and making large profits for unscrupulous ship owners, assuming the ship arrived safely. Wrecks were commonplace in the days of sail and fire at sea was an expected hazard. On the other hand, collisions were few, as both ships needed a stout wind to gain sufficient momentum to do each other damage. A ship becalmed in fog was delayed but unlikely to come to much harm.

With the coming of steam navigation, all this changed. On the one hand, steam propulsion brought greater safety because it provided masters with the power to deal with gales, and allowed them to steer away from treacherous lee shores. But this advantage was outweighed by the risk of running out of coal and the chance, in those radarless days, of colliding at night or in fog with other vessels.

The fogs of the north Atlantic are the worst in the world. For weeks on end in spring and autumn fog drives around the Grand Banks of Newfoundland, reducing visibility to nil and covering the whole sea with a sound-deadening blanket which has to be experienced to be believed. It was in conditions such as these that the early steamship accidents occurred, although many skippers at first tended to treat fog more as a nuisance than anything else. The famous first Commodore of the Cunard Line, Captain Judkins, dealt with fog by increasing to full speed and tearing through the murk flat out, because, as he put it 'you are then sooner out of it'.

It was in such a fog on the 27 September 1854 that the Collins liner *Arctic* encountered trouble in the notorious Grand Banks area. The *Arctic*'s Captain Luce, an old-time sailing skipper whom Collins had transferred to his steamers, pressed on at full speed. On board with him he had Edward Collins' wife, Mary Ann, his youngest son Henry, and his only daughter, who was named after her mother. In addition, Captain Luce had brought along his handicapped eight-year-old son, who was making the voyage for the sake of his health. Among others was the Duc de Gramont, newly-appointed French Ambassador in Washington. As the *Arctic* tore on at full speed through the fog, one of the lookouts suddenly spotted a vessel ahead in the mist. His shout of warning almost coincided with the collision.

The *Arctic* had been holed by the iron hull of the French steamer *Vesta*, whose shadow rapidly disappeared in the mist astern. The damage to the *Arctic* was lethal, although Captain Luce at first believed that she had suffered only superficially. Then his engineers reported to him that three holes had been made in her way below the water line.

The run for safety was over before it begun. The ship began to heel, and

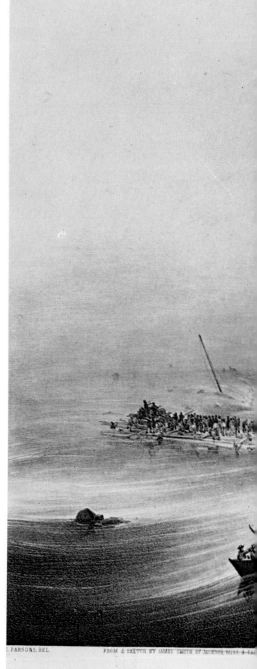

*The Collins liner
Arctic sinking off the
Grand Banks of
Newfoundland after a
collision with the
French steamer
Vesta. The Arctic's
wooden hull filled
rapidly and 346
people drowned,
including the wife and
two children of
Edward Collins,
founder of the line.*

OF THE U.S.M. STEAM SHIP "ARCTIC"

OFF CAPE RACE WEDNESDAY SEPTEMBER 27TH 1854.

slowed to a stop. As many women and children as possible were assigned to the lifeboats. At this point, it must be recorded that the crew of the *Arctic* panicked, entered all but one of the lifeboats and left the sinking vessel. The Chief Engineer crammed as many of the women and children as he could into the remaining lifeboat and got clear but, when the *Arctic* foundered about four hours after the collision, only forty-five people survived of the 391 on board when the ship left Liverpool. Only two of the lifeboats ever reached the shore and among those drowned were Mrs. Collins and her two children and the French Ambassador. Captain Luce went down with his ship, clasping his crippled son, but wreckage from the main paddle box fell on the child, killing him instantly. Luce himself managed to keep clear and clung to the wreckage for two days, before he was picked up by the Cunard liner *Cambria*.

The *Arctic* disaster was the first blow of many suffered by the Collins Line which had, up until the ship's loss, established an ascendancy over its Cunard rivals. Sixteen months later, the ocean dealt another blow to Collins, this time using one of nature's most ferocious weapons – the iceberg. On 23 January 1856, the *Pacific*, sister ship of the *Arctic* and the first ever to cross the Western Ocean in nine days, sailed from Liverpool with forty-five passengers and 141 crew on board. The shortage of paying customers indicates the reluctance of voyagers in those days to commit themselves to a winter journey. The *Pacific* passed the Liverpool Bar and headed down St. George's Channel, bound for New York.

Following her to sea a day or so later, went the new Cunarder *Persia*, built in stout iron and setting out to beat the Collins ship and restore something of her owner's reputation. Five days out from Liverpool, the *Persia* ran straight into an icefield while steaming at 11 knots. Before any avoiding action could be taken the liner's bow was bent over, 16 feet of plating torn from her starboard side and the starboard paddle box and

paddles badly damaged. Undoubtedly, her iron hull saved the *Persia*. Down by the head, she was still able to make New York. On arrival Captain Judkins found that the *Pacific* had not preceded him and the Collins liner was never heard from again. Her owners could only assume that, whereas the *Persia*'s iron hull had stood firm, the *Pacific*'s wooden one had failed, and that their ship and all her people had perished in the icefield.

Some years later, another of the early liners suffered an encounter with ice: this was the Guion Blue Riband challenger *Arizona*. Homeward-bound from New York in November 1879 under the command of Captain Jones, the *Arizona* encountered a typical Newfoundland Grand Banks fog. Captain Jones reduced his speed but, even so, little could be seen in the murk and the *Arizona* ran head on into an iceberg, causing her bows to collapse. Passengers in the smoking room were thrown to the floor and those who recovered quickly enough ran on deck to see the *Arizona*'s bow literally buried in an iceberg towering sixty feet above the ship. In fact, the bow had been crushed right back to the ship's collision bulkhead, but she remained sound and Captain Jones brought her safely into St. Johns, Newfoundland. There, a temporary wooden bow was fitted and the *Arizona* raced back to Liverpool in a startling six days, seventeen hours and thirteen minutes! Captain Jones was held responsible for the accident and his Master's Certificate suspended, but the *Arizona*'s reputation soared. The travelling public argued that any ship that could collide head-on with an iceberg and survive was strong enough to do it again, and more people booked on the *Arizona* than on any other liner at the time. For once, misfortune had proved commercially profitable.

The 1870s provided the most lethal wreck in nineteenth-century steam navigation when the White Star liner *Atlantic* was wrecked on the Newfoundland coast, taking 585 people down with her. The *Atlantic* in her early career had been a money-spinner for the White Star Line

Two examples of ships that survived head-on collisions. The Guion Line Blue Riband holder Arizona *in St. Johns, Newfoundland, after ramming an iceberg in November 1879. The incident impressed the travelling public as clear evidence of the ship's strength.*

The bows of the Italian migrant carrier Florida *after she had rammed and sunk the White Star liner* Republic *in fog south of Nantucket on 23 January 1909. Thirty feet of the* Florida's *bow has been crushed down to five feet.*

and she was two years old when on 20 March 1873 she left Queenstown for New York with 942 people on board and enough coal to keep her engines turning for fifteen days. From the start the voyage was anything but a happy one and later there were to be accusations that the ship's officers were 'carousing the whole voyage'. Eleven days out, the *Atlantic* was still over 400 miles from the American coast, when Chief Engineer Foxley told her captain that the coal supply was down to 127 tons. Captain Williams decided to alter course for Halifax to take on more coal. As the ship neared the Newfoundland coast, a rough sea got up and Captain Williams, before turning in, left orders to be called before the ship approached the entry to the port of Halifax.

But no one woke him and the next sound that Williams heard was at 0315 hours when his ship crashed on to the rocks of Prospect Cape, 20 miles from Halifax. Williams immediately dashed on deck to take control of a crew in which panic was rapidly spreading. The sea swept away all the ship's boats on the port side and only one of the starboard lifeboats had been launched when the *Atlantic* fell over on her side and sank.

Her bow and part of the rigging remained above water and the helpless survivors clung to this last refuge until taken off by fishing boats out of Halifax. The subsequent court of enquiry suspended Captain Williams' certificate although it recognized his brave conduct after the accident. Nevertheless, the tragedy of the *Atlantic* was a definite setback to the emerging White Star Line. To the astonishment of the line's officials the court went on to add that, in its opinion, the *Atlantic* had sailed short of coal and the officials spent years in legal battles trying to reverse the decision. As for Captain Williams, he was back on the bridge of a White Star liner within two years of the *Atlantic* affair!

The Inman Line, too, had its share of disaster during the 1870s. Its *City of New York* went aground on Daunts Rock outside Queenstown and was a total loss. Then the *Glasgow* caught fire at sea and had to be abandoned. But the worst tragedy suffered by the Liverpool line was in January 1870, when their *City of Boston* left Halifax for Liverpool on the 28th of that month. She had come up from her namesake port in Massachusetts and had 177 people on board. After leav-

ing the Nova Scotian port, nothing more was ever heard of the *City of Boston* and her name was added to the role of ships that regularly sailed out to oblivion.

Even in these later years of the 20th century, when technology reigns supreme, it is not unknown for ships to vanish, equipped though they may be with every advanced radio and navigation aid. In the 19th century, such affairs were commonplace and even the early years of the new century can offer an example of a large passenger liner, newly built and well-officered, disappearing almost within sight of land. This was the Blue Anchor liner *Waratah*, an emigrant carrier on the service between the United Kingdom and Australia via the Cape. Homeward bound in June 1909, the *Waratah* called at Durban and then set out on the short coastal run round to Capetown. During the early stages of what was almost an overnight run, she was sighted by several other ships. Then she vanished and was never heard from again. The *Waratah* story remains one of the great unsolved mysteries of the sea or of any other element for that matter. A modern, well-found liner, equipped with radio, is overwhelmed by some unknown force within a hundred miles of

port on a crowded sea-lane and not a single oar, spar or mangled body is left to tell anxious would-be rescuers that a great ship has passed!

The coming of wireless telegraphy at the turn of the century changed for all time the safety of humanity at sea. Yet, as with many other social or technological advances destined to change the course of history, it took a decade or more to awaken the world as a whole to the possibilities of Marconi's invention. The persistent little Italian had required all the determination he possessed – and it was plenty – to convince the Victorian financial world that his ideas were worth backing. He had first sent messages over the air from the American liner *Saint Paul* in 1898 and the first permanent installation of a Marconi set was in the *Kaiser Wilhelm der Grosse* in 1900. But progress was slow and many shipowners were reluctant to pay out good money on new-fangled apparatus which was only good for keeping up with the stock exchange prices when the ship was near enough to land to come into range of receiving stations. If shipping companies were slow off the mark, for once the world's naval staffs were quick to realize that here was a new

The American Line's St. Paul, 11,629 tons, from which Marconi first sent wireless messages from ship to shore in 1898.

element in naval warfare and the wireless played a decisive but largely unrecognized part in the victory of the Japanese fleet over the Russians in May 1905.

Then there occurred, in the space of just over three years, two shipwrecks which placed Marconi and his box of tricks in the forefront of world attention. The first was sensational enough. The second was apocalyptic, involving the largest ship in the world in history's most famous shipping disaster.

Both incidents involved ships of the White Star Line, then at the height of its fame and power and probably considered by most people as the world's foremost nautical enterprise. As dawn broke over Nantucket on Saturday 23 January 1909, the White Star liner *Republic*, 15,400 tons, was just south-west of the island, outward-bound for the Mediterranean. sunshine with 400 passengers. Sunshine was

the thing that the *Republic* needed most, for she was cautiously feeling her way through a typical Western Ocean fog. At 5.40 a.m. a horrified lookout on the liner saw another ship loom up in the mist, and before any avoiding action was possible, crash into the port side of the *Republic*.

Then the other ship backed off and vanished as quickly as she had come. With her engine-room flooded and a dangerous list to port, the White Star liner was alone in fog, disabled and in danger of foundering. Now it was that Marconi's invention made its first call for help from a ship at sea. The ship that had dealt the *Republic* her mortal blow was the Italian liner *Florida*, bound for New York with 900 emigrants, mostly refugees from the recent Messina earthquake. As she sliced into the *Republic*, killing three passengers, she also collapsed the radio shack where Marconi operator Jack Binns was sound

Typical watertight doors similar to those fitted on the Titanic. *The doors were hydraulically-operated and are shown open and partially closed. The collision damage to the* Titanic *was so extensive that the watertight doors failed to save the ship.*

asleep. Then all electric power failed as the sea rushed into the generator rooms. But Binns quickly rigged his set to emergency batteries and sent off his now historic signal: 'CQD. All stations Distress. *Republic* rammed by unknown steamship 26 miles southwest of Nantucket. Badly in need of assistance.' The signals were at once picked up by the French Line's *La Touraine* and White Star's own *Baltic* who both altered course for the scene. Although the *Baltic* was only 64 miles away, she actually steamed 200, because of the fog, before she came up with the *Republic* at 7 p.m. that same evening. By then, the *Florida* had groped her way back to the disabled *Republic* and the two captains decided that all the

Republic's people would be safer aboard the Italian vessel. Later they were transferred again, when the *Baltic* arrived.

Only then did Jack Binns quit his post on the *Republic*, which sank the following day as attempts were made to tow her into New York. Binns became a hero overnight and was given a typical New York welcome parade. Thousands turned up to watch the *Baltic* arrive with the rescued. More significantly, there was an instant flurry of press editorials demanding that all large ships should be fitted with the new invention of the wireless.

Just two months to the day after the accident to the *Republic*, the White Star Line's greatest ship yet was laid down at the Belfast yard of Harland and Wolff.

The Titanic *leaving Belfast Lough for her sea trials on 2 April 1912. In thirteen days time, she would be at the bottom of the Atlantic, after history's greatest peacetime sea disaster.*

The new monster was the second of the three large ships, twice as big as any existing rival, that the White Star chairman Bruce Ismay planned to give his fleet absolute domination of the north Atlantic passenger services. On these ships there would be unheard-of luxury for those who could afford it and, below, more spartan but no less profitable, there would be spaces available for the emigrants who still hurried in their thousands to the elusive promises of the New World. Ismay gave his ships names that suited their role in the creation of big things. The first was the *Olympic* and the new ship too would express the irresistible progress of her owner's ambitions. She was to be the *Titanic*. She was ready on time for her

The great hull of the Titanic *safely afloat after her launch at Belfast on 31 May 1911. The ship's plating is only single skin as the double bottom was not extended up the sides of the hull. If it had been, she may well have survived the collision with the iceberg.*

maiden voyage in April 1912 and she came down from Belfast to Southampton to embark a first-class passenger list that exactly matched the world's largest and most luxurious ship. There was the great Colonel John Jacob Astor, head of America's greatest financial clan and founder of the Waldorf-Astoria, the banker Ben Guggenheim, Charles Hays, President of the Grand Trunk Railroad, and his opposite number from the Pennsylvania, John B. Thayer. The roster of millionaires continued with Isodor Straus, founder of the great New York department store Macey's, who was returning from a European tour with his wife. From the world of politics came Major Archibald Butt, military aide to President Taft, and London society was represented by the Countess of Rothes and Sir Cosmo and Lady Duff-Gordon.

Somehow, in that spring of 1912 the maiden voyage of the *Titanic* took on a social significance far beyond the simple fact of crossing from Europe to America. Most of her millionaire passengers were

The 46,329-ton Titanic *in profile. The damage occurred on the starboard side and extended from the bow for over 300 feet to well aft of the bridge and opened up six watertight compartments to the sea.*

regular travellers of the ocean route and not a few reckoned the great liners as a natural extension of their offices in Wall Street or their mansions in Tuxedo Park or Oyster Bay. It was just such a market that Ismay and J. P. Morgan had forseen for their three great ships, and now Ismay himself came along to join a group which included passengers, business associates and, in many cases, good personal friends.

If Ismay was the acknowledged prime chieftain of passenger shipping at that time, he had with him on the *Titanic* two men who probably knew more about the building and sailing of superliners than any other pair outside Cunard. Thomas Andrews was managing director of Harland and Wolff and he was on board with a team from the yard to see the new ship safely into service so, as he said, 'to see she does the old firm credit when she sails'. In 1912 it was the role of company heads to lead from the front.

The *Titanic* had grown largely from Andrews' own drawing board and now he was putting the final touches to his master-piece by making the maiden voyage himself and seeking out the minutest flaw in the liner's perfection. He took a great deal of trouble to get the hatracks right. The other half of this outstanding maritime duo was the *Titanic*'s master, Captain Edward J. Smith. 'E.J.', as he was known, was everyone's idea of the typical express liner captain. Tall, bearded and imposing, his very presence exuded authority and he probably had more experience of hand-ling big liners than any other man afloat. Smith had commissioned the *Adriatic* and the *Olympic*, both designed by Andrews, and he therefore knew the Belfast engineer well.

The story of the maiden voyage of the *Titanic* is also the record of the most famous wreck in history. The *Titanic* remains the largest ship ever to sink on the high seas in peacetime and her death roll has never been exceeded, except in war years. It is not surprising that, as the years roll by, a forest of sturdy legend has grown up around the ship and that simple incidents, unremarked at the time they occurred, later assumed an altogether ominous and foreboding importance in the light of the disaster. The first of these omens gained substance at noon on 10 April, minutes after the *Titanic* cast off from her berth in Southampton's Ocean Dock. To gain the seaward chan-nel, the liner had to pass several other ships tied up nearby. One of these was the American Line's 10,500-ton *New York*, formerly the Inman Blue Riband chal-lenger. As the *Titanic* passed, the suction from the large ship's hull dragged the *New York* from her moorings and hauled her stern-first out into the fairway. As snapping hawsers scattered quayside watchers, Captain Smith cut his engines and the danger passed. The *Titanic* sailed on to Cherbourg and then Queenstown. There, as the last tender went ashore, at 1400 hours on Thursday 11 April, the magnificent *Titanic* headed out into the Atlantic with 1,316 passengers and 892 crew.

'E.J.' himself had contributed to the *Titanic*'s reputation for unsinkability.

Captain Edward J. Smith, master of the Titanic. *He was the highest paid sailor of his time.*

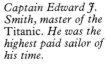

'Modern shipbuilding has gone beyond that,' he declared to a reporter who quizzed him on the likelihood of disaster occurring to one of his previous White Star commands, the *Adriatic*. True, the *Titanic* was as safe as standards then demanded. She was divided into fifteen watertight compartments and she could float with any two of them flooded. She could even float with all four of her bow compartments open to the sea, and no one could foresee a worse situation than that. So confident were the authorities of shipping safety that a ship the size of the *Titanic* was required to carry lifeboats for 1,178 persons, the maximum demanded by law. But that same law also allowed the ship to leave port with 2,208 people on board.

Until the following Sunday, the trip proceeded without incident. Captain Smith was breaking the ship in gently and it was only on this day that the engineers worked the *Titanic* up to 22 knots. The weather was fair, the food and company excellent and even in the steerage high spirits prevailed among the hundreds of emigrants on board.

During 14 April, however, the *Titanic*'s Marconi operators began to receive warnings of icebergs ahead in the general direction of the liner's track. Both the Cunard *Caronia* and White Star's own *Baltic* – she of the *Republic* rescue – sent warnings and Captain Smith thought the latter to be of such importance that he showed it to Bruce Ismay, and later posted it in the chartroom. Then, just as the day was ending, and the watch changing at 2200 hours, Jack Phillips, the *Titanic*'s senior radio man, took down the following signal from the Atlantic Transport Line's *Mesaba*:

'In lat. 42 to 41 deg., 25 mins. north, long. 49 deg. west to 50 deg., 30 mins. west, saw much pack ice and great number heavy icebergs, also field ice.'

Jack Phillips was an overworked man. All day he had labored to pass passengers' private messages ashore via the station on Cape Race and he still had a large backlog. So he carried on with his work and did not pass the *Mesaba*'s

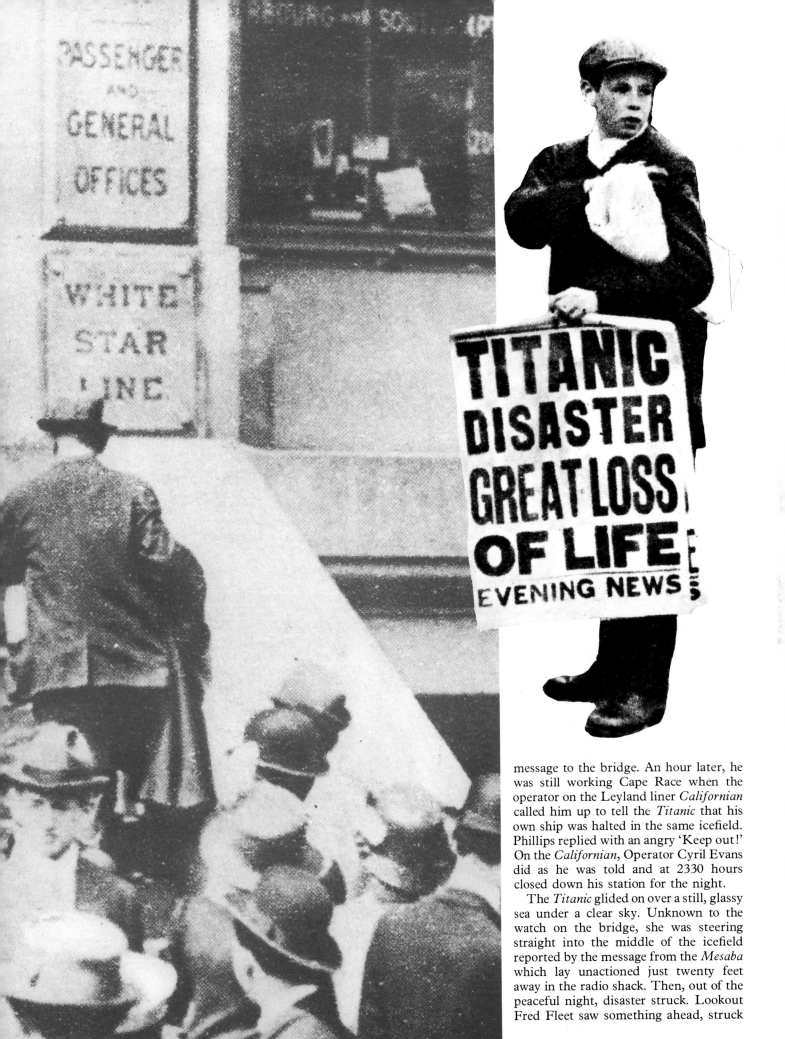

message to the bridge. An hour later, he was still working Cape Race when the operator on the Leyland liner *Californian* called him up to tell the *Titanic* that his own ship was halted in the same icefield. Phillips replied with an angry 'Keep out!' On the *Californian*, Operator Cyril Evans did as he was told and at 2330 hours closed down his station for the night.

The *Titanic* glided on over a still, glassy sea under a clear sky. Unknown to the watch on the bridge, she was steering straight into the middle of the icefield reported by the message from the *Mesaba* which lay unactioned just twenty feet away in the radio shack. Then, out of the peaceful night, disaster struck. Lookout Fred Fleet saw something ahead, struck

135

the crow's-nest warning bell three times and telephoned the bridge, 'Iceberg dead ahead!' It was 2340 hours. Second Officer Murdoch, Officer of the Watch, ordered the helm over, but it was too late. Fleet watched in horror as the *Titanic* ran her starboard side athwart a gigantic berg which towered above her decks and then vanished astern. The big ship gave a slight shudder which went unheeded by large numbers of her sleeping passengers. Some got up to see what had happened but most went back to bed. For Captain Smith and his officers there could be no such complacency. Water was pouring in below and a quick survey of the damage showed that over 300 feet of the hull aft of the starboard bow had been opened up to the sea. On the bridge, Smith conferred with Andrews who quietly explained that, with the first six compartments open to the sea, there was no chance that the ship could remain afloat. Orders were given to uncover the lifeboats and put off the women and children.

The events that followed combined all the elements of classical tragedy and heroism. Phillips' first radio call for assistance flashed into the air when the new day was but five minutes old and six ships answered immediately, including the *Olympic*, the stricken liner's sister. Nearest of all was the Cunarder *Carpathia*, only 58 miles away and ambling along at 11 knots, outward-bound for Gibraltar and Naples. Captain Rostron ordered his ship about and told his radio operator Harold Cottam: 'Tell him we're coming along as quick as we can'. That night Rostron ran his ship at speeds he had no idea she was capable of, as he dodged berg after berg to race to the *Titanic*'s last known position. But he was too late.

Slowly the great ship sank by the head while the boats were launched. In death, Andrews rose to heroic stature as he helped many women to escape the ship he had designed and built. He and all his Belfast men were lost, as were Chief Engineer Bell and his entire staff who remained at their tasks until the last, maintaining electrical power that illuminated the ship

almost until her final plunge. With them went Astor and all the millionaires, while old Mrs. Straus refused all attempts to separate her from her husband, saying that they had started together and would finish in like manner.

Throughout the sinking, many of the passengers on the *Titanic* saw a ship, fully lit, about ten miles off. Later evidence led the British Court of Enquiry to identify this mysterious and inactive onlooker as the luckless *Californian* and her unfortunate skipper Captain William Lord had to endure the stigma for the remainder of his long career. The matter remains controversial to this day.

But some men did enter the boats. Stokers, migrants and even some who were invited by the ship's officers, like Sir Cosmo Duff-Gordon. But one who made his own decision was Bruce Ismay. In a snap, last-moment decision that he must have regretted all his remaining life, the White Star chairman entered a lifeboat. Once ashore he became the target of the more virulent American newspapers who argued, not without justice, that Ismay owed his place in the boat to any one of the hundreds of steerage women and children who died in the terrible disaster.

At 0220 hours on 15 April, the end came. The *Titanic*, weighed down at the head by hundreds of tons of sea water, stood on her beam ends. She remained so for a minute or so, then as watchers in the

lifeboats looked on in awe, she slowly slid below the surface, taking 1,503 victims with her. The *Carpathia* arrived just two hours later and Rostron's crew hauled the dazed survivors aboard. Then he headed for New York, where several newspapers had believed the already circulating rumors and ran headlines announcing that all aboard *Titanic* had been saved. When the truth came out, a worldwide storm of criticism struck the unfortunate White Star Line.

The *Titanic* disaster was a turning point in maritime history. Laws were passed that made the provision of lifeboats for all on board compulsory and an international iceberg patrol was set up to track down the great 'growlers' and warn all ships of their presence. Marconi shares soared as the little Italian's invention again bathed in the spotlight of world publicity. Severe standards of hull construction were laid down by national governments of the larger maritime powers and the *Olympic* was sent back to Belfast for her bulkheads to be raised and extra boats taken aboard. The fact that

the survival rate among first-class male passengers was higher than third-class children escaped comment in the contemporary press, something impossible today. Subsequent official enquiries by the American and British Governments were surprisingly lenient in the strictures they passed on the ship's owners and officers. It cannot be denied that the *Titanic*, fully loaded, was navigated at high speed into an area where icebergs had been reported not once, but a number of times. Whatever else this action may be called, it certainly must at least be condemned as careless seamanship.

Much about the *Titanic* disaster remains unexplained. Although the boats held spaces for 1,178 people, only 706 were aboard the eighteen boats picked up by the *Carpathia*. The argument about the *Californian* has gone on undiminished since the night the *Titanic* went down. It can be argued that the ship seen by the *Titanic* was not the *Californian* but that there was a ship in sight was beyond doubt. Too many witnesses observed its distant light for there to be any question about the matter.

The legends remain. The most famous concerns the tune played by the ship's band as the liner went down. Every survivor questioned by the press in the succeeding weeks claimed that Bandmaster Hartley and his musicians struck up the hymn 'Nearer, my God, to Thee', as the boat deck dipped under. It was im-

possible to check with the performers as Hartley and his entire band perished, but trained observers who later wrote accounts of the disaster claimed that the last tune was the Episcopal hymn 'Autumn'. Any argument about the tune pales before the heroism of the musicians who played on until the end.

After the *Titanic* disaster, the White Star Line was never the same again. It still had twenty years or so of life left to it, but the memory of the disaster lingered on and there can have been few passengers to step aboard a White Star vessel in the succeeding years who did not give the *Titanic* at least a passing thought. Although it took twenty years for the White Star Line to die, Bruce Ismay was destroyed overnight. In less than a year he had resigned all his directorships and retired to live as a recluse in Ireland. He died during the thirties.

So cataclysmic was the loss of the *Titanic* that any succeeding tragedies were automatically judged by public opinion on 'Titanic' standards. Two years later, a disaster of almost equal proportions occurred in the St. Lawrence river and passed almost unnoticed in comparison to the publicity given to the ill-fated *Titanic*.

Compared to the *Titanic*, the Canadian Pacific liner *Empress of Ireland* could hardly be described as a floating palace. She was a 15,000-ton ship engaged in the routine passenger trade between Canada

and the United Kingdom and she had left Quebec for Liverpool on the afternoon of 28 May 1914. Her passenger list did not contain the famous and privileged names that were commonplace on the New York run. The ship was under the command of Captain Kendall – the captor of Dr. Crippen – and that night he dropped his pilot at the usual spot off Father Point. Soon afterwards, fog closed in and the *Empress of Ireland* collided with a small Norwegian freighter called *Storstad*. The *Storstad* was carrying 10,000 tons of coal and she punched a hole in the *Empress* along the water line. The Canadian Pacific ship sunk as fast as any vessel on record. She went down in less than fifteen minutes, turning over on her beam ends and sinking before many of her passengers could even get on deck, let alone into the lifeboats; 1,106 people were drowned. The tragedy for some reason never assumed the legendary proportions of the *Titanic* and, even today, very few people are aware of it. Perhaps the outbreak of war three months later and the huge casualty lists from armies on both sides of the Western Front drove such small matters as shipping losses out of the public mind.

The years between the two World Wars were marked by a series of bad ship fires – the other main danger to seafarers. During that time three big French liners were all lost through fire and although the death toll was negligible, the

TRAGEDY IN THE ST. LAWRENCE

The Canadian Pacific liner *Empress of Ireland* leaving the pierhead at Liverpool at the start of one of her voyages to Canada. She was homeward-bound from Quebec with 1,370 passengers when the accident took place and went down in less than fifteen minutes, taking 1,106 people with her. Below: the *Empress* heeled so rapidly that it was impossible to launch the boats on the port side.

Collier STORSTAD
which rammed the
IRELAND
showing damaged
Bows

Above: the damaged
bows of the Norwegian
collier Storstad *after*
she had rammed and
sunk the Empress of
Ireland.
Right: contemporary
artist's impression of
how the collision
occurred.

LAKE ONTARIO

QUEBEC CITY

CRANE I?

HARE I?

QUEBEC
PROVINCE

RIVER St LAWRENCE

I? VERTE

------- 30 miles across St Lawrence -------

BIG I?

BARNABY I?

RIMOUSKI

Wireless message
received at
Father Point

Wireless
Aerial

Wireless Station   Lighthouse

FATHER POINT

C.P.R. LINER
"EMPRESS of
IRELAND"
14,500 Tons

SOUTH SHORE
OF
St LAWRENCE

POINT OF COLLISION
AMIDSHIPS

THE NORWEGIAN
COLLIER
STORSTAD
6000 Tons

From Nova Scotia

5

Left: the end of the Morro Castle: sightseers throng the foreshore at Asbury Park, New Jersey, to watch the final hours of the burning liner, aground and abandoned on 8 September 1934. The captain of the Morro Castle later faced charges under American maritime laws.
Right: at the end of the Second World War, the German Blue Riband holder Europa was ceded to France as the Liberté. Before she could be refitted, the Liberté broke from her moorings in a gale on 12 December 1946 and struck the wreck of the Paris, which had lain in Le Havre throughout the war years. The Liberté sank and did not go into service until four years later.

world was left in no doubt that a big modern passenger ship could burn with an intensity that defied efforts to extinguish the flames. During the 1920s there was a serious fire on the *Berengaria* and the German liner *Europa* was seriously damaged while fitting out for her maiden voyage in 1929. Both these ships were saved, although the *Europa* had to be scuttled to put out the fire.

It was not until well after the Second World War that very large passenger ships began to appear on routes other than the north Atlantic passage between Europe and the United States and Canada.

There was, nevertheless, one exception which occurred in the late 1920s on the south Atlantic route from Bordeaux to Rio and the Plate estuary, operated by the French consortium, Compagnie Sud-Atlantique.

In 1928, this company ordered a giant ship, the largest ever built for South American passenger services, from the Penhoet yard at Saint Nazaire. Because of the restrictions imposed by the narrow channel of the Gironde estuary, the new ship was restricted in length to 745 feet for a gross tonnage of 42,000. Her consequent large beam and three low funnels gave her a squat appearance, and

there was little improvement after the height of her funnels was raised during her first annual refit. Her complete lack of shear added to her ugly looks.

The ship was named *L'Atlantique* and given the same high standards of accommodation as the north Atlantic record-holders. Her first-class dining saloon was on the same lavish scale as those later installed in the *Normandie* and the *France*. She had a central avenue 450 feet long and 20 feet high with rows of luxury shops on either side, which soon became known to crew and passengers alike as 'La rue de la Paix'. Too large to come up the Gironde to Bordeaux itself, *L'Atlantique* used the Pauillac quay downstream from the city, and her maiden sailing in September 1931 brought new standards of luxury to the south Atlantic. But the big French ship did not last long. On 4 January 1933, on her way from Bordeaux to Le Havre under the command of Captain Schoofs, for annual dry docking, fire broke out at 0330 hours in cabin 232 on E deck, 22 miles west of the Guernsey coast in position 49° 30′ N., 3° 17′ W. The fire spread rapidly and the ship was abandoned by her skeleton crew as dawn broke. Nineteen lives were lost. Well ablaze, the wreck drifted out of control

towards the Dorset coast off Portland Bill, from which she was plainly visible on 5 January, while a French naval vessel, the minelayer *Pollux*, stood by ready to sink her if need be.

Taken in tow at last, on 6 January, *L'Atlantique* was brought into Cherbourg. There followed a long lawsuit between the underwriters and the owners. The latter wished to declare *L'Atlantique* a constructive total loss and claimed her complete value (£2 millions), although the Belfast yard of Harland and Wolff gave an estimate for complete repair which was much lower (£1,250,000) than the owner's claim. The wreck lay at Cherbourg until the Sud-Atlantique obtained judgement in their own favor, after which she was sold to the Scottish shipbreakers Smith and Houston Ltd. in February 1936 and was broken up at Port Glasgow.

The fire on *L'Atlantique* confirmed that a large ship, crammed with highly combustible fittings consisting mainly of wood and carpeting in often inaccessible locations, was a travelling fire hazard of the most dangerous proportions. The difficulties facing ship's fire-fighters were two-fold. First, they found the centre of the fire was in some inaccessible spot and,

once there, the use of too much water would cause the vessel to heel over as her stability was destroyed.

In the year that followed the fire on *L'Atlantique*, another deadly blaze caught the attention of the world's press. The *Morro Castle* was a medium-size ship of 11,520 tons that traded between New York and Havana in Cuba. She was typical of a number of American liners of similar size that operated from east coast ports to the Caribbean. In the early hours of 8 September 1934, fire was discovered in the ship's library. The *Morro Castle* was then 20 miles south of Scotland Lighthouse which stands at the entrance of New York harbor; she was carrying 318 passengers and 240 crew. The fire soon took hold and spread rapidly through the ship. There was a fatal delay in sending radio messages for help and only eight of the ship's twelve lifeboats were able to be launched. Strong swimmers among those on board took to the water and some actually made it to the New Jersey coast, but 180 people died in the flames. The *Morro Castle* disaster shook American opinion; several of the crew were charged with neglect of duty and American safety standards considerably improved as a result of the incident.

The 1930s closed with another spectacular fire. On 19 April 1939 the Transat liner *Paris*, 34,500 tons, was lying in Le Havre when fire broke out on board. The *Paris* had survived a previous fire in August 1929, but this time her luck ran out. So much water was pumped on board that the ship heeled over to port and eventually sank alongside the berth. Lying there on her beam, her great bulk prevented the *Normandie* from leaving the dry dock just beyond the *Paris* where she was under repair, and the mast of the *Paris* had to be cut off in order to let the other ship out. The outbreak of war six months later meant that the burnt-out wreck lay untouched until 1946 when the work necessary to remove it was started. By then, the *Europa* had wrecked herself on the wreck of the *Paris*!

The Second World War brought new developments in navigational aids, among which was radar. This brilliant invention allowed ships to scan the seas around them and mark the position of any approaching vessel or obstacle, even in the thickest fog on the darkest night. Nevertheless, as the years have gone on, ships are still wrecked and one or two still vanish without trace. A number of passenger liners have been lost by fire and

the most spectacular post-war blaze has already been described – the destruction of the *Queen Elizabeth* in January 1972.

Two post-war incidents underlined that, despite modern ingenuity and technical invention, the oceans are still dangerous and great vigilance is still demanded of all who sail on them. On 19 December 1963, the Greek liner *Lakonia*, 20,314 tons, left Southampton on a Christmas cruise to the sunshine of the Canary Islands. She carried 651 passengers and 390 crew under Captain Mathios Zarbis. Three days later the *Lakonia* was 100 miles north of Madeira when fire broke out in one of her public rooms and spread rapidly throughout her upper decks. The *Lakonia* was an old ship, built before the Second World War for a Dutch company, and the fire proved too much for her crew to control. Rescue ships hurried to the scene and 500 survivors were taken off by the Argentine liner *Salta* alone. Nevertheless, ninety-one people were killed and another sixty-four reported missing, before the fire was eventually allowed to burn out. The incident provoked much criticism of Greek liners and their standards of crew discipline and training, and the Greek government introduced tough new laws

Left: the Swedish-America line Stockholm *after her collision with the* Andrea Doria. *Her bows are completely crushed and have torn part of the Italian ship away with them – including an unlucky passenger.*
Below: survivors from the Andrea Doria *being helped by medical staff aboard the veteran French liner* Ile de France.

to prevent a repetition. The *Lakonia* herself sank a week later while under tow to the Rock of Gibraltar.

The second incident has proved to be the classic collision in modern shipping history. In July 1956, the Italian liner *Andrea Doria*, flagship of the Italia Line, with every radar aid possible and built to the highest standards, was inbound for New York and nearing the end of her voyage. As she navigated through patchy fog her radar picked up a bleep indicating that another ship was approaching. This turned out to be the Swedish liner *Stockholm*, another fully equipped modern liner. Both ships were aware of the other's presence and yet, because of faulty interpretation of their respective positions and a misunderstanding of the anti-collision

rules, the *Stockholm* crashed into the *Andrea Doria*, killing forty-three people and slicing the *Andrea* open through seven of her eleven decks.

The Italian ship immediately heeled to such a degree that only half her lifeboats could be launched. Forty-four miles away the old French liner *Ile de France* answered the *Andrea*'s SOS and raced to the scene, while the *Stockholm* herself lowered boats to pick up survivors. There followed a controversial scene where, to the disgust of the *Stockholm*'s crew, they discovered that such boats that had been launched from the Italian ship were crammed with stewards and other crew members. Had it not been for the arrival of the *Ile de France* and the prompt action of the *Stockholm*, many of

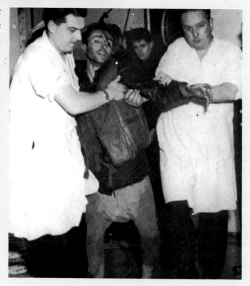

The Andrea Doria *soon after her collision with the* Stockholm. *She has heeled sharply to starboard and all her port lifeboats remain in place, incapable of being launched, due to the angle of the ship. Even so, some of the forward promenade deck windows have been removed in the hope of launching the boats. The* Andrea Doria *is the largest liner to be lost at sea since the Second World War.*

The final plunge of the Andrea Doria. *Lifebuoys and deckchairs are strewn on the sea as the pride of Italy's merchant fleet goes to the bottom of the Atlantic.*

the *Andrea*'s 1,134 passengers might well have been left to their own devices. Finally, after all passengers had been transferred, the *Andrea Doria* was left adrift and listing in a calm sea. At 1000 hours the following morning, with a circle of ships around her, the largest Italian liner of the day turned over and went under.

Almost within reach of the American coast and equipped with everything that could be required to prevent such a disaster, the *Andrea Doria* had gone to the bottom. It was a sad and salutary lesson. There followed a long battle in the courts between the two shipping lines, but a victory in the courts could do nothing to replace the lives of the dead or rebuild the ship.

CHAPTER FIVE

HIGH SEAS SOCIETY

Life on board the liners

On board the Ile de France, *chic French ship of the twenties.*

'I NAME THIS SHIP *Queen Elizabeth the Second*. May God bless her and all who sail in her.' With these words, Queen Elizabeth II of Great Britain on 20 September 1967 named the latest in a long succession of liners to enjoy royal patronage. The big ships had long been dubbed 'Queens of the Seas' or 'Monarchs of the Ocean' by company publicity staff and journalists alike. Their size and splendor went well with the trappings of monarchy. One complemented the other, and almost all the big European ships had royal sponsorship at their launching ceremonies. Even in republican France, these occasions were graced by the President himself. President Lebrun was there in October 1932 when his wife named the *Normandie*, and no less a person than President Charles de Gaulle watched the *France* go down the same slipway in 1960. President and ship had their own peculiar contribution to make to the greater glory of France.

The roll of European monarchs was severely shortened by the First World War, and after the Second it almost disappeared entirely. But in the Age of Kings, crowned heads took a benign and often political interest in national progress towards ever bigger liners.

Doyen of the business was Kaiser Wilhelm II of Germany. Behind his impetuous arrogance, Wilhelm was a true patriot with a clear understanding of maritime power. He needed little encouragement to become interested in German shipping and saw the prestigious advantages to be gained from possession of the world's largest ships. In the years before 1914, he was present at nearly all the launching ceremonies in Germany, including those of the *Kaiser Wilhelm der Grosse* and the *Imperator*.

It was in fact the Kaiser's grandfather, Prince Albert, consort to Britain's Queen Victoria, who started the business of royal involvement with steamship launching ceremonies. On 19 July 1843, he left London's Paddington Station at the crack of dawn for Bristol and the naming ceremony of the *Great Britain*. The

THE RO
TO THE G

The Orient Line's Ophir *was specially commissioned into the Navy as a Royal Yacht to take the Prince and Princess of Wales on a Royal Tour in the east in 1901.*

engine of the Royal Train was in charge of I. K. Brunel himself, who brought the royal party safely over his 100-mile track in a creditable time of just over two hours. Thirty thousand onlookers awaited the ceremony (it was hardly a launching, as the *Great Britain* was already afloat in her dock) and all drew a sharp breath as Mrs. Miles, mother of a local M.P., missed the ship altogether as she threw the now customary bottle of champagne! Prince Albert now became the man for the occasion. Seizing a second bottle, providentially held in reserve, he cracked it squarely on the iron bow as the *Great Britain* passed by. 'The stupendous progeny of the genius of Mr. Brunel' was safely afloat and properly named.

Two years later, in January 1845, Albert was to bring his beloved Victoria to see the *Great Britain* as she lay in the Thames, and the little Queen, who was an indefatigable diarist, recorded that all was to her satisfaction and 'most tasteful'. It was the beginning of a long association with liners by Queens of the House of Windsor. All of them would launch ships in their day, and two named the largest passenger ships ever built. There is a delightful and probably apocryphal story that the original Cunard choice of name for the *Queen Mary* was *Victoria*. Answering a query by King George V about the proposed name, a high Cunard official rather pompously replied that the chosen name was that of the most gracious lady ever to sit on the throne of England. King George was delighted and said he would tell her the moment he got home!

True story or not, *Queen Mary* the ship became, and the royal lady was there to name her (with a bottle of Madeira!), as indeed her daughter-in-law was for the naming ceremony of the *Queen Elizabeth* three years later.

Elsewhere, King Victor Emmanuel accompanied his wife to the launch of the Italian *Rex*, and Queen Wilhelmina saw the Holland America's *Nieuw Amsterdam* safely afloat in 1938. Curiously enough, the White Star Line, no stranger to bally-hoo when it suited, never encour-

147

Royal patronage for Holland America, as Queen Wilhelmina of the Netherlands names and launches the Nieuw Amsterdam *on 10 August 1937. The ship was a popular favorite on the Atlantic for over thirty years.*

BUILDING
S.S. NIEUW AMSTERDAM
33000 TONS

aged a formal naming ceremony for its ships. Those grandees who turned up for the occasion watched the serious business of launching a liner without such fripperies as beautiful ladies and bottles of champagne. But one lady slipped through the net. When the *Baltic* was launched on 21 November 1903, the Belfast press recorded on the following day that she had been named by Miss Julia Neilson, star of the show 'Sweet Nell of Old Drury', then playing at the Belfast Opera House. One can only speculate on the reason for this departure from White Star practice!

Royal patronage was (and is) important to a shipping company, and the attendance of a monarch at the inaugural moments of a new ship was good for prestige and profit. But launching day was only a beginning, and throughout a ship's career, her owners would go to some lengths to obtain the favor of the famous and the wealthy. Not that this was difficult to obtain in the golden Edwardian days or in the twenties and thirties. The ocean voyage had long since ceased to be a dubious adventure, where a certain risk obtained. It was now rather a travelling, four-day-long party, where old friends would be encountered and new ones made. Sailing day at New York, Southampton, Le Havre or Bremen always produced some celebrity or other, and reports of arrivals or departures were a regular feature in newspapers of the day. The wandering press photographer of the thirties spent his days lingering around Pier 90, Manhattan or South-

ampton's Ocean Terminal, hoping to land a shot of a commuting film star that all-important hour ahead of his rivals on another paper.

The arrival of talkies in 1928 boosted the film industry to a global enterprise. For a few pennies or cents ordinary people everywhere could, for a brief hour, escape from an often humdrum life, into a world of high adventure and romance. The stars themselves, though, were real enough and every detail of their lives and loves was eagerly scanned by thousands of followers. The liners, with their opulence and splendor, were ideal transportation for the stars, who found them a ready-made stage in real life.

The *Ile de France* was a popular favorite with the stars. Tallulah Bankhead, Constance Bennett, Gloria Swanson, Jeannette Macdonald and Grace Moore all used her at one time or another. With the stars came the legends and stories, now so delightful in the telling, yet so hard to confirm as true. For instance, which Hollywood lovely one gala night on the *Ile* stopped the show by appearing in a ravishing new example of *haute couture*, with one breast fully exposed? And was it really true that Captain Blancart once stopped the great ship in mid-ocean during a gaie and pumped out fuel oil to calm the raging sea so that Isadora Duncan could dance at a Seamen's Charity Gala without risking a broken leg? It is doubtful, but it makes a marvellous story. Blancart was prepared to admit later that he did once alter course to hold the *Ile* steady in high seas while the great dancer Argentina performed, and also that he summoned up the ship's orchestra to play outside the bathroom of a French beauty who confessed her love for music when bathing.

The late Maurice Chevalier was an *Ile* regular, who sometimes challenged the skills of the chef with recipes all his own, including his mother's mutton stew; which needed two days in the cooking.

Musicians also seem to have taken a liking to the *Ile de France*. The great Toscanini would travel on no other ship

ILE DE FRANCE
Chic pleasure palace of the twenties

Widely popular with anyone at all
conscious of fashion and social style
during the twenties and thirties, the
Ile de France attracted both the
famous and the scandalous to her
passenger lists. On board the ship
reputed to have the most romantic
atmosphere of any liner on the
Atlantic, they could relax in the
lounges decorated by Jeanniot,
Bouchard and Saupique or dine in
the Patout dining room which was
embellished with three shades of
Pyrenean marble. The liner even
featured in Noel Coward's song,
'These Foolish Things'!

*World heavyweight
boxing champion,
Primo Carnera
presides over a boxing
match on board.*

Couples swirl gracefully around the ballroom.

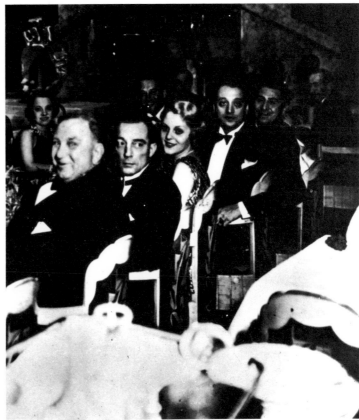

The captain's table, the social center of life on board. This group includes Buster Keaton and Maurice Chevalier.

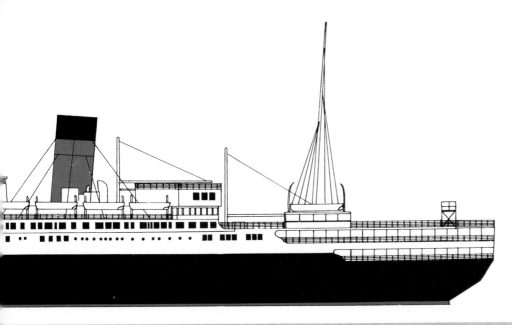

*Enrico Caruso,
the great Italian
tenor (and party!)
on board the* Kaiser
Wilhelm II *in
September 1903.*

if he could avoid it. He once arrived in New York on a Wednesday, conducted the Philharmonic on Friday evening, and sailed out, again on the *Ile*, on Saturday. During one voyage on the *Ile*, Toscanini had his first meeting with the young Menuhin. Other *Ile* passengers were Sigmund Romberg of 'Desert Song' fame, Heifetz, Chaliapin, Paderewski and Rachmaninoff, the latter often playing his own works to an admiring audience of fellow passengers.

The novelist Ernest Hemingway told a story of a meeting at sea which, one feels, could only have happened on the *Ile de France*. In the days before his work became recognized worldwide, Hemingway, who was travelling second class on the *Ile*, gatecrashed the first-class dining room in order to dine with a friend, who loaned him the necessary dinner jacket, without which he would not get past the entrance to the dining room. As the two men dined, a beautiful woman in a white dress appeared at the head of the grand staircase. It was Marlene Dietrich, who slowly made her way to a table where twelve people rose to greet her. Dietrich protested that she could not make thirteen at table, but the alert Hemingway dashed in to offer his services as fourteenth man!

A gala atmosphere pervaded the first-class accommodation of the big Atlantic liner in those days. On other seas, a more staid society, going out for a spell in the service of the Government of India, or setting out for a business deal in South Africa, would quietly co-exist for days on end with perhaps no more excitement than a weekly concert. But on the Atlantic, gaiety was the order of the day and ingenious purser's staffs created all sorts of games and diversions for their paying guests.

Deck games were always popular, particularly shuffleboard and deck tennis, and there would be tug-of-war, pillow fighting and other delights in a daily programme of ship's sports. Keep-fit fanatics abounded, guided by the fiction that five days at sea left one a

When the film industry finally aspired to the status of global enterprise after the coming of the talkies, commuting film stars became familiar figures on the great ships of the north Atlantic. Top left: Clara Bow and her husband, Rex Bell. Below left: Sally Eilers Gibson. Center: Marlene Dietrich. Top right: Pola Negri. Below left: Douglas Fairbanks Sr. and Jr.

155

*Crossing the Line!
King Neptune and
court come aboard a
Holland America
liner, and the tradi-
tional ducking takes
place of all those
crossing the line for
the first time.*

flabby, incapable and unfit human being.
True, the prodigious meals on board
found out many a delicate stomach, but
seldom laid anyone low for more than a
day or so. The author well remembers
emerging on the promenade deck of the
Queen Mary early on the first morning he
spent aboard and having to avoid pairs of
frantic walkers making circuit after circuit of the liner's quarter-mile promenade.
Ashore, most of these promenaders would
scarcely have gone out of their way to
cross the road to buy an evening paper!
Others would explain, if asked and often
if not, that they were 'getting their sea
legs', for the spectre of sea-sickness lay
not far from the thoughts of all but the
most seasoned travellers.

Companies had an ambivalent approach to 'mal-de-mer'. On the one hand,
they were prepared to spend good money
on the anti-rolling tanks of the good
Doctor Frahm (sometimes these only
speeded up the roll) or fit bilge keels,
while on the other, they treated what
could be a wretched experience with a
casual nonchalance, expressing surprise
that anyone at all could be sick on one of
their vessels. Frankly, sea-sickness had
to be endured stoically by those who
suffered from it. As the ship headed into
a gale, out came ropes to be rigged across
lounges and promenades, and in the
dining rooms tablecloths were dampened

LIFE ON DECK

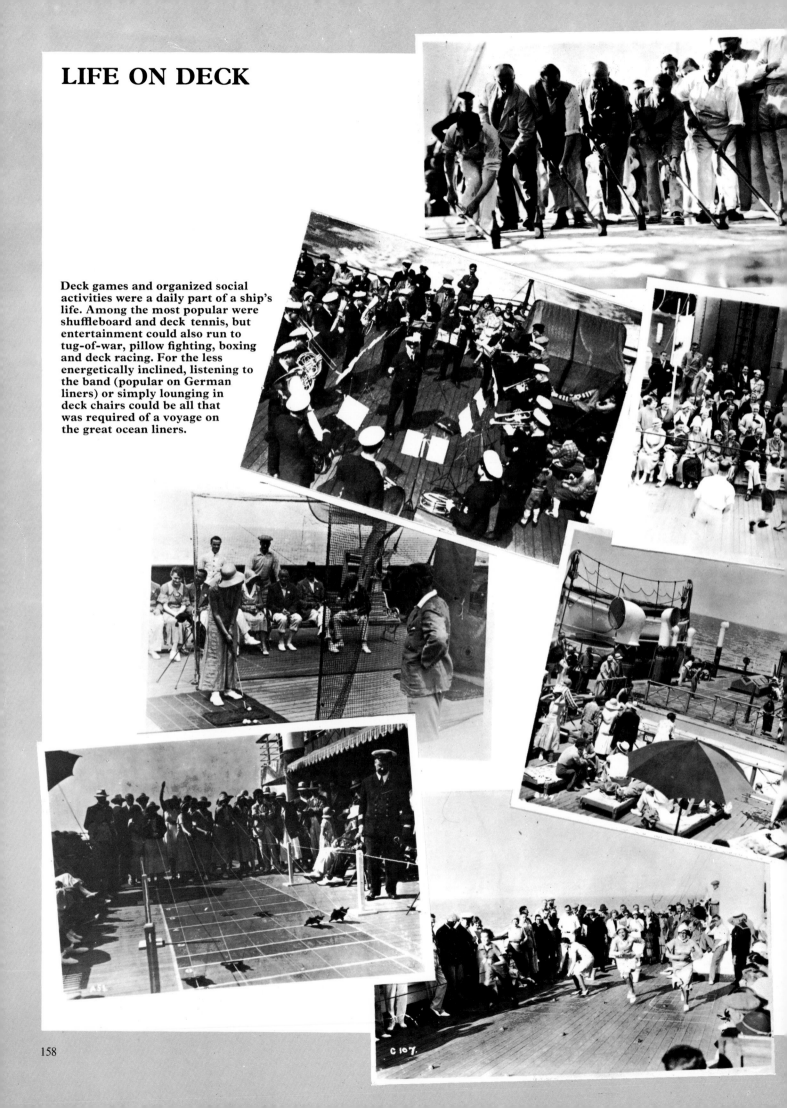

Deck games and organized social activities were a daily part of a ship's life. Among the most popular were shuffleboard and deck tennis, but entertainment could also run to tug-of-war, pillow fighting, boxing and deck racing. For the less energetically inclined, listening to the band (popular on German liners) or simply lounging in deck chairs could be all that was required of a voyage on the great ocean liners.

to keep the crockery in place. The Hapag liner *Deutschland* once lost $2,000-worth of porcelain for lack of this simple and elementary precaution.

And so they came, those fabulous passengers of the superships, arriving on sailing day with at least twenty valises or trunks, and watched over by attentive servants and maids who often made up a good quarter of the passenger list. Among them would be a sprinkling of millionaires who all possessed not only unlimited wealth but style as well. Benjamin Guggenheim, millionaire member of the New York banking family, who lost his life in the *Titanic*, attempted to persuade his valet to leave him and try to reach safety. The man refused, so he and Guggenheim went below and later re-

turned on deck in full evening dress. 'We have dressed in our best and are prepared to go down like gentlemen', said the banker, and both men perished when the ship foundered.

The *Titanic* disaster allows us to measure the wealth that might be represented on a single Atlantic crossing. Colonel Astor had $2,500 in his wallet when his body was recovered from the sea and his fortune was estimated at $150 millions. To that could be added $95 millions for Guggenheim, $50 millions for old Mr. Straus and a mere $40 millions for Bruce Ismay.

Where such wealth abounded, there came also the criminal classes, alert for the pickings that could be had in plenty from the less astute voyagers, though sel-

dom from the millionaires. Card sharpers abounded, and despite the warning notices posted in all ships' public rooms, there was always the sucker, often rich and travelling alone, who would take up the invitation to play cards with a stranger. Most victims were the subject of a polished performance, and later, when they had lost several thousand dollars, they were mostly too ashamed to admit their gullibility and the crooks went free. Another 'situation' was the beautiful 'wife' of a fellow-passenger who claimed her husband was always drunk and she was always alone. The affair would proceed until the final dénouement in the 'wife's' cabin with the sudden appearance of the 'husband' and the subsequent blackmail of the would-be lover. Again

SEASICKNESS
TRAINSICKNESS

Positively prevented
and cured by

Mothersill's
SEASICK REMEDY

Left: 'Dressing under difficulties' is the laconic caption of this print of life aboard a P. & O. liner of the 1880s.
Right: many patent medicines were on offer to sufferers from seasickness. Most companies treated it as something to be endured stoically, and only in recent years has the scourge been overcome by the drug dramamine and injections.

the criminals were seldom caught, but the ship's crew saw most of the game and came to recognize crooked travellers who appeared too frequently. Given the slightest opportunity, liner captains leaned heavily on the card sharpers and gamblers, and Sir Arthur Rostron, Commodore for Cunard on both the *Mauretania* and *Berengaria*, once claimed that he had reduced gamblers' gains by at least $50,000 while in command of those two ships.

One criminal who was caught, however, was Dr. Crippen, and the manner of his arrest deserves a place in both forensic and maritime history. Harvey Hawley Crippen was an American who had settled in London and worked as a medical technician. Although he had no

degree, he persisted in the use of the title 'Doctor' and it is as Doctor Crippen that he holds his unique place in the history of criminology. Crippen was a small, mild-mannered individual, and his huge moustache did little to hide this. He suffered daily humiliation from his extrovert wife, a lady who had pretensions to a music-hall career under the name Belle Elmore. The situation developed all the elements of tragedy when the little doctor met and fell in love with Ethel Le Neve who was working as his secretary. As Belle's excesses grew, so Crippen's passion for Ethel increased until at last in January 1910 he poisoned his wife, dismembered her and hid part of her body below the cellar floor of their London home. Crippen reassured sus-

picious neighbours that his wife was on a visit to America, and later announced that she had died there.

Ethel Le Neve, meanwhile, had moved in with the doctor. Nemesis for Crippen now appeared on the scene in the form of Chief Inspector Dew of Scotland Yard, who had been called in by stage friends of Mrs. Crippen who did not believe the Doctor's story. Dew himself was half inclined to believe Crippen when he confessed that his description of Belle's death was phoney and that he had made it up to hide the fact that she had left him. Dew went back to the Yard and took a weekend to think things over. On the Monday, Dew went to Crippen's house to resume his enquiries and found it empty. Crippen and Ethel Le Neve had vanished and there, after a search, Dew found the putrefied remains of Belle. An enquiry that had started as nothing more than a missing person check-out now became a full-scale murder hunt and an international hue and cry was raised. A warrant for the arrest of the fleeing pair was issued on 16 July 1910.

Four days later, the Canadian Pacific liner *Montrose* sailed from Antwerp on a routine trip to the St. Lawrence ports. Once at sea, Captain Henry Kendall began to take more than a passing interest in one of his passengers, a Mr. John Robinson who was travelling with his teenage son. Allowing for natural affection between father and boy, the pair appeared to Kendall to be on strangely intimate terms. Kendall, aware of the hunt for Crippen, did some detective work of his own and was soon convinced that he had discovered the fugitives. He radioed his suspicions to Liverpool and Dew left the following day on the White Star's *Laurentic*, a fast ship scheduled to arrive in Quebec two days ahead of the *Montrose*. By now, the press had the story and the world watched in excitement as the two ships headed west. The chase ended when Dew arrested Crippen and his lover on board the *Montrose*, as she halted off Quebec's Father Point. He brought them back in the *Megantic* to face

The Daily Mirror

THE MORNING JOURNAL WITH THE SECOND LARGEST NET SALE

No. 2,108.   Registered at the G. P. O. as a Newspaper.   FRIDAY, JULY 29, 1910   One Halfpenny.

"CRIPPEN BEYOND A DOUBT"—YESTERDAY'S WIRELESS MESSAGE FROM CAPTAIN KENDALL, OF THE STEAMER MONTROSE.

Mr. Llewellyn Jones, Marconi operator on the Montrose.

Inspector Dew, who will bring back the fugitives.

Captain Kendall, of the steamer Montrose.

The steamer Montrose, on board which are Crippen and Miss Le Neve.

Miss Le Neve, who accompanies "Dr." Crippen dressed as a boy.

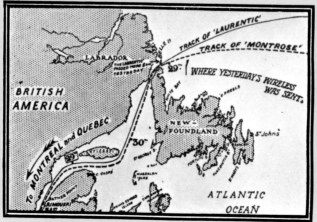

Chart showing the positions of the Montrose and the Laurentic.

"Dr." Crippen, who is now definitely said to be on the Montrose.

A wireless message was received yesterday from Captain Kendall, of the steamer Montrose, which stated that the identity of the suspected passengers on the ship had been established "beyond a doubt." It will be remembered that two passengers who entered their names on embarking on the Montrose at Antwerp as "Mr. Robinson and son" were suspected as being "Dr." Crippen and Miss Le Neve, for whom a world-wide search was being made, and as a result of a wireless message from the captain Inspector Dew, of Scotland Yard, was sent to overtake the suspects on the faster steamer Laurentic.

162

The tragic Gay
Gibson, a British
stage actress, pictured
in happier days. She
was murdered on
board a Union Castle
liner in October 1947
by a steward, who
disposed of her body
by pushing it over-
board through her
cabin porthole.

trial. The court acquitted Ethel, but Crippen was hanged, and the world marvelled anew at the genius of Marconi who had made it all possible.

Murder on the high seas is a rare event in modern times and usually confined to quarrels among crew members. A celebrated case occurred in October 1947, however, aboard the Union Castle liner *Durban Castle* which stirred the interest of press and public because the victim was a glamorous young actress named Gay Gibson. She was returning to England after a stage tour in South Africa when she was found to be missing from her cabin. A search of the ship revealed nothing and although the *Durban Castle* retraced her course, no sign of the girl was found and she was assumed lost overboard. Then a steward reported that he had seen one of his colleagues in Gay's cabin late on the night of her disappearance and his story led to an amazing trial at Winchester Assizes. Without a body in the case, the accused, a steward called James Camb, had a good chance that his story that Gay died from natural causes while he made love to her would be believed. Camb said that in panic he had pushed the actress's body out through a porthole. The court, however, chose to accept the prosecution's version that Camb had murdered her and he was convicted.

Fortunately such violent incidents occur infrequently in the social history of the liner trade. The main concern of the big companies continued through the years to be the fame and fortune of their clients and the profit that could be made therefrom. This preoccupation with the plutocracy had developed, as we have seen, in the 1890s. Before that, one passenger was very like another. But records of ocean crossings by celebrities of Victorian times survive and none is more fascinating than the journey to America made by Oscar Wilde in 1882.

Wilde was then at the height of his defiance of Victorian social convention. He had come over in the Guion Line's *Arizona*, she of the famous iceberg inci-

Where is the red-head from cabin 126?

By FRED REDMAN

With the Union Castle liner Durban Castle, homeward-bound from South Africa, was sailing up the African coast, first class passengers held a dance.

At midnight, pretty, auburn-haired Eileen (Gay) Gibson, 21, a London actress who had been appearing in shows in South Africa, was laughing in the arms of a partner.

Next morning her cabin No. 126, was found to be empty. There were blood stains on the pillow.

The ship turned back for a fruitless hunt for a missing girl. When they reached Cowes Road...

Miss Gibson was in her cabin soon...
"I hea...
ing th...
sounde...
being...
notice...
Pa...
her...
that...
looki...
turn...
spirits...
boarded...
"never...
than she...
Gay, w...
given as N...

...r. Hampstead, went to South ...ht. Africa in the Capetown Castle a few months ago to appear with Eric Boon, ...rmer British light-weight ...ing champion, in "The ...lden Boy," at Johannes-...g. Her parents, who ...t to South Africa with ..., are still there.

Yesterday detectives re-...oved the door of Cabin ...6, sealed since Miss Gib-...on's disappearance, and handed it over to finger-print experts.

The police learned that the girl's nightdress is miss-ing. The porthole of her cabin was wide open at ...wn.

...black evening gown, ...e she wore at the ...was hanging on a

...en strap on one of ...son's dance shoes, ...the cabin, may be ...tant clue.

*A Cunard poster
promoting the early
voyages of the* Queen
Mary. *A trip to
America gave the
ordinary citizen a
chance of rubbing
shoulders with the
famous.*

*The Duke and
Duchess of Windsor,
regular Atlantic
commuters, pause for
the* Queen Mary's
photographer.

dent. Outrageously dressed, and with his hair touching his shoulders, Wilde was met by a group of reporters at what must have been one of Manhattan's first press conferences. Here an alert reporter caught on his note pad the supreme epigram of Wilde's caustic career as a commentator. As the customs officers made their routine check, one asked Wilde the stock question, 'Anything to declare?'. 'Nothing but my genius,' came the reply, and Wilde went ashore to carry out a year long tour of the States arranged for him by Rupert D'Oyly Carte.

On his return to New York, Wilde hastened to welcome Lily Langtry, the Jersey Lily of the Victorian music hall and close friend – some said too close – of Edward, Prince of Wales. Lily had also crossed in the *Arizona* and complained of rats in her cabin. Thirty years earlier, another songstress, Jenny Lind, 'the Swedish nightingale', had also crossed to New York to appear for showman Phineas T. Barnum. She travelled in Collins' *Atlantic* (no doubt without a rat in sight), and was welcomed by 30,000 New Yorkers whipped up by Barnum for the occasion.

It was not until after the First World War that the liners could claim the regular patronage of a class that until then had avoided them. While tycoons, businessmen, actresses, writers, politicians and aristocrats all patronized the big ships, royalty still used royal yachts or warships for any ocean travel they might be disposed to undertake.

The nearest King Edward VII of Great Britain ever came to a liner was a visit he paid to the White Star *Teutonic* in company with the Kaiser in August 1889, and Wilhelm, although a liner enthusiast as we have seen, always used his yacht *Hohenzollern* when he went to sea, although he once spent a night on the *Imperator*. King George V likewise visited liners, including the *Majestic* in August 1922, but it was left to his eldest son, the late Duke of Windsor, as Prince of Wales, to break with convention.

The Prince let it be known that he had

Contented card players gamble a few cents while the ship's orchestra plays in the background. Many travellers became the victims of shipboard cardsharps.

booked as a private passenger on the *Berengaria* to cross to New York during August 1924. From the moment of the announcement, the ship was booked solid by Americans, anxious to rub shoulders with the heir to the throne of the British Empire. Cunard ordered up a band to greet His Royal Highness and caused consternation among the ship's officers by issuing instructions that swords would be worn! Down from London came orders that the Prince insisted on no special treatment and both band and swords were hastily cancelled. The Prince put his words into action by slipping aboard the *Berengaria* from a launch in the Solent, leaving a frustrated army of photographers milling about the quay at Southampton. Travelling with the Prince were Lord and Lady Louis Mountbatten and that night the royal party occupied the Imperial Suite, the only previous royal occupant of which had been the Kaiser eleven years before, when the *Berengaria* had started life as the *Imperator*.

The following day was Sunday and the usual service was conducted by the Captain, William Irvine, in the lounge. As a diversion on Sundays, divine service was obligatory in Victorian times, but by the twenties, the Captain was lucky if one in ten of his passengers turned up. That particular Sunday the lounge was packed and passengers had to scramble for chairs. The Prince duly attended the service, having breakfasted in his suite, thereby avoiding hundreds of unusually early risers who had waited for him in the dining room. Once the voyage was well under way, the Prince indulged in all the usual shipboard activities that awaited the average first-class passenger. He led a tug-of-war team (it lost!), he was knocked off his perch in the pillow fight, and in the potato race suffered the ignominy of losing his potato! The ship's orchestra soon discovered to their cost that Prince Edward was an ardent dancer, who could, if necessary, dance away into the small hours. He usually danced with ladies in his party, but occasionally the

Gambling of another kind. Passengers queue to buy tickets for the tote on the ship's daily run. The tote was run on all liners and in all classes.

world would explode for an American girl as the next King of England swung her across the dance floor planned so carefully by Mewès years before. In this way did the future Duke of Windsor make his first acquaintance with America, a country he came to love, and one of whose daughters he would eventually marry.

All that was an age away from the *Berengaria*'s great black hull and brightly-lit lounges, as she carried the Prince westward in that summer of 1924. Altogether he was away three months and returned to Britain in the *Olympic* at the end of October. The Prince of Wales was known as a trend-setter (he could make a fortune overnight for men's fashion designers by simply adopting a new style) and Royal visits thereafter tended to favor the liners as transport. King George VI used the Canadian Pacific's *Empress of Australia* for his visit to Canada and the United States in 1939, and returned in style in the huge *Empress of Britain*. His daughter used the Shaw, Savill and Albion liner *Gothic* in similar fashion for her post-Coronation tour of the Commonwealth in 1953.

With Royal patronage and interest, passengers needed little encouragement to obey the publicity men when they offered the temptation to 'live like a king',

Above: the apotheosis of European civilization afloat, an artist's impression of the magnificent Grand Salon of the Normandie.
Left: an exciting evening in the Normandie's Grand Salon, as immaculately dressed passengers applaud the climax of a cabaret act.
Right: a well-tried part of entertainment on board: entrants for the ship's Fancy Dress Competition.

168

as one brochure put it. The liners employed an army of chefs, cooks, butchers, bakers, pastrymen, pantrymen, cellarmen, winewaiters, waiters and stewards to ensure that they did. No bill of first-class fare was complete without a dozen or so courses, and on each course a choice of five or six items. The organization required to supply these floating palaces was complex and immense, and whole businesses in Southampton and New York flourished on ship supplies. The quantities of food were prodigious. Here are just a few examples from a single voyage of the *Queen Mary*: 106 tons of beef, 25 tons of potatoes, 18 tons of vegetables, 1 ton of sausages, $\frac{1}{2}$ ton of tea, 70,000 eggs, 1,000 jars of jam, $2\frac{1}{2}$ tons

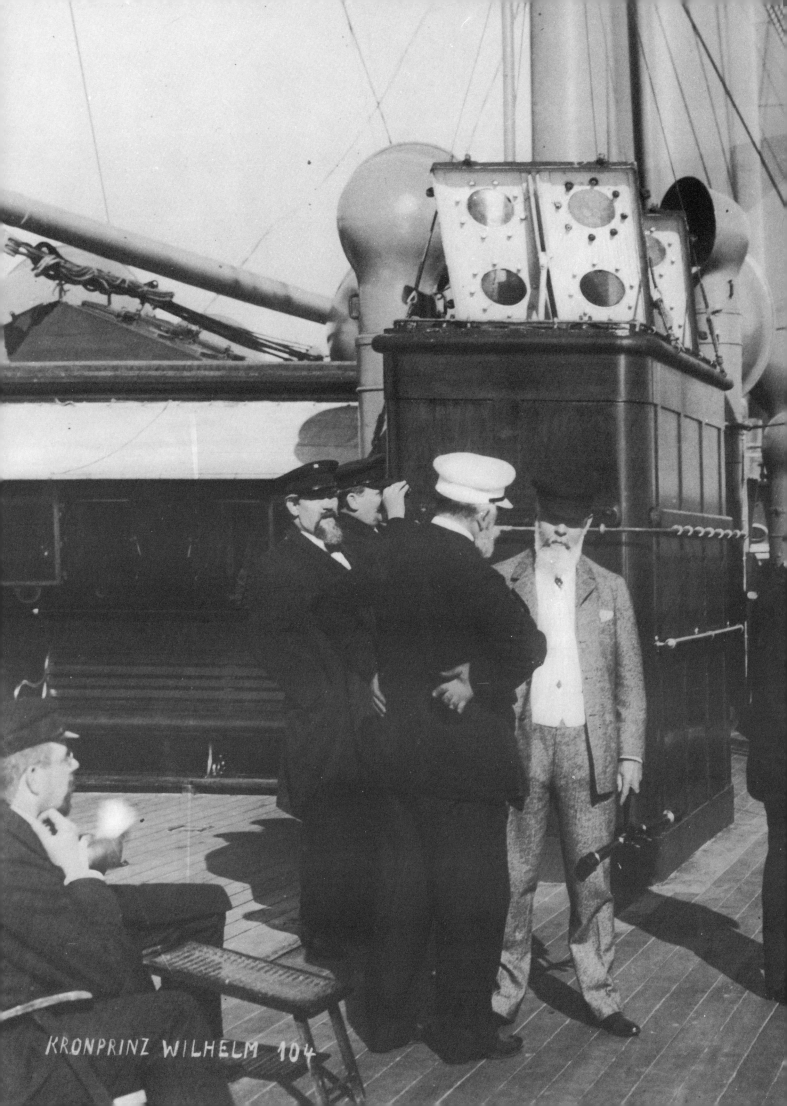

KRONPRINZ WILHELM 104

A fine day aboard the Kronprinz Wilhelm. Gentlemen chat amiably while others study the horizon. A rising sea could rapidly change this tranquility into chaotic waveswept motion.

of bacon, 160 gallons of salad oil, 2,000 gallons of ice cream, 9 tons of fish, $\frac{1}{2}$ ton of bananas, 500lb of smoked salmon and $4\frac{1}{2}$ tons of lamb. The bar stocks were equally formidable. Shipboard drinking was a way of life for some passengers and booze was cheap, being exempt from the excise duty imposed ashore. A big liner would consume 2,500 bottles of whisky, 6,000 bottles each of beer and lager, 3,000 bottles of table wine, 48,000 bottles of mineral waters and 2,500 bottles of stout per voyage. While consuming these items the passengers also got through 1,500,000 cigarettes and 15,000 cigars.

The catering superintendents had an unequalled opportunity to buy the best, as the full choice of British, French and American foods was available to them on every voyage. Florida melons, paté de foie, fresh fillet beef: the best from each country, bought at source at the keenest prices and converted into a vast array of dishes that seems almost revolting in its magnitude. But passengers expected it and the companies obliged. Right up to the end of the Atlantic service, the traditional Cunard menu survived on the *Queens*, and one must wonder just how much food ended, not inside the passenger, but in the trash can during those last unprofitable years.

The crew needed to sustain the high life of the liners was never less than 1,000 men on the big ships, the majority of them on the catering and stewarding side. The ship could be navigated by ten officers and forty or so seamen, and the engineering staff was only fifty or so after the conversion to oil fuel in the early 1920s. Before that, a big liner would need 200 stokers or firemen to toil in filthy conditions feeding the ceaseless demands of the furnaces. Few passengers ever saw the degrading black hole that was the stokehold or the tough men that shovelled 500 tons of coal a day to keep the ship alive and make the glittering fantasy of the first class a reality. The conditions below bred a line of tough, ruthless men who took some handling, and engineer officers learned early in their careers to

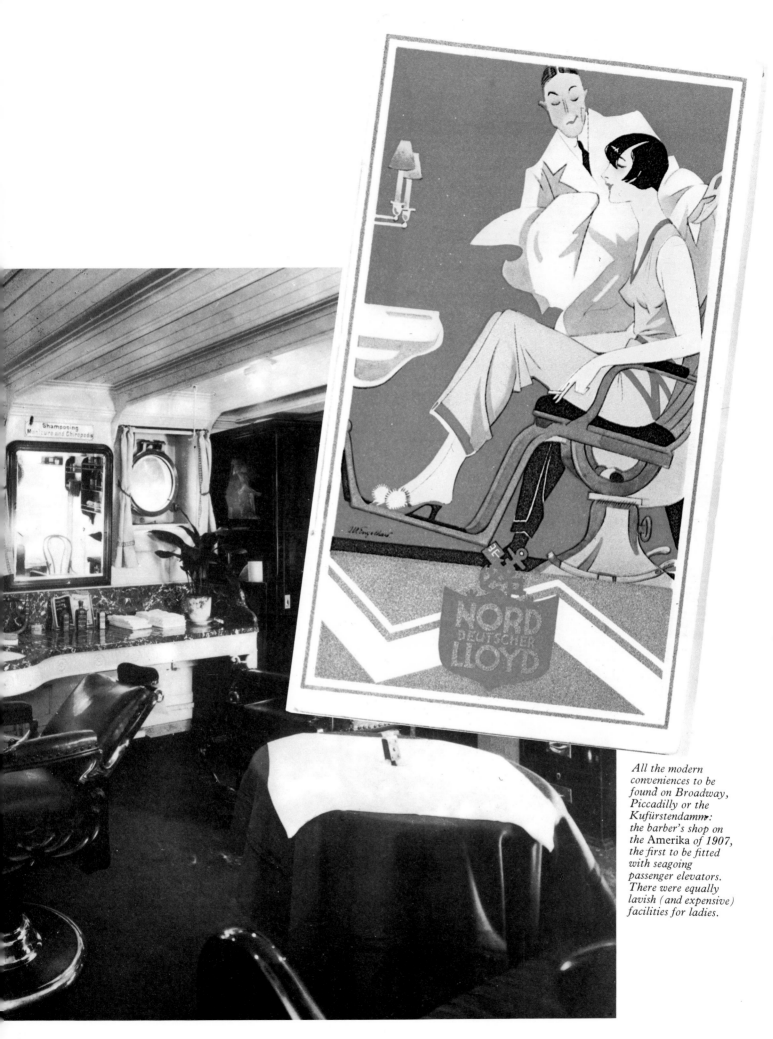

All the modern
conveniences to be
found on Broadway,
Piccadilly or the
Kufürstendamm:
the barber's shop on
the Amerika of 1907,
the first to be fitted
with seagoing
passenger elevators.
There were equally
lavish (and expensive)
facilities for ladies.

173

Left: the stoke-hold of the *France* of 1912. Conditions were dirty, unhealthy and tough and the 'black gang' had a reputation for aggressive independence in ports worldwide. The hell that was the stokehold was never seen by passengers and it vanished when the conversion to oil fuel took place.

Right: on the bridge of the Vaterland. Although protected ahead, the wheelhouse is still open to the elements.

Below: up for a breath of fresh air, the entire engine room staff of the Cunarder *Campania* pose for the camera. Most of the men in the picture are the firemen needed to feed the ship's furnaces.

keep clear of a stokehold brawl. One White Star engineer, known for his hard-driving attitude, is alleged to have ended up in the furnace for his pains.·

When oil fuel was introduced, the conditions in the furnace rooms took on an unnatural cleanliness overnight and stokers appeared in sparkling white overalls. They were a new breed, and the old 'black gang' disappeared, leaving a legend behind it of drunken brawls and a fear of the Liverpool Irish that persists in ports to this day.

To command such crews, and rise to the top of their profession, captains needed to summon all the qualities of command and personality that they possessed. The top captains were impressive men. Most of them started in sail, and were from humble backgrounds. Yet men like Sir James Charles of Cunard or Bertram Hayes of White Star could handle a Duke at one end of the social scale and a recalcitrant stoker at the other with equal aplomb. E. J. Smith of the *Titanic* was an Atlantic patriarch who sometimes took his pet wolfhounds to sea with him. Millionaires found that all their money could not buy a place at his table if 'E.J.' took a dislike to them. Another great captain of the thirties was Arthur Rostron.

These men had an almost emotional involvement with their ships. Sir James Charles, due to retire, collapsed and died as he brought the *Aquitania* into Southampton for the last time. His doctor claimed that the emotional stress of the occasion had been too much. A few years later, Sir James Netley died in his cabin on the same ship. Again it was his last voyage, and Sir Edward Britten of the *Queen Mary* died in the same way on a morning his ship was due to sail from New York. Other great captains still remembered by Atlantic veterans were Hans Ruser of the *Imperator*, Pierre Thoreux of the *Normandie*, and the giant Harry Manning, who presided over the United States Lines for many years.

When the liners were in their heyday, the captains were international figures.

Essential to the running of any liner, the ship's printing shop produced daily hundreds of items such as newspapers, programmes, menu cards and routine orders. This busy scene is on the Europa.

Left: bell-boys on the Queen Elizabeth *line up for inspection. Many crew members started their careers in this way.*
Right: the kitchen staff were a vital element in making a liner's reputation and companies treated their galley employees well. These cooks are aboard the German liner Auguste Victoria.

The senior captain in the Cunard service invariably received a knighthood from the King. They were men who were familiar with fame and the famous. And the famous came in hordes. Even after the war an average Atlantic voyage might have a passenger list which contained the Duke and Duchess of Windsor, Spencer Tracey, Bing Crosby, Ernest Hemingway, and even the *QE2* on one of her early voyages could claim the prophets of the new age, the Beatles!

Gradually, however, the famous, the aristocrats, the businessmen, transferred their loyalties to the airlines and the liners retired from the scene. The airliner could cross the ocean in a fraction of the time that the liners needed and in the late 20th century, time meant money and money was everything. It was the search for wealth and the power of wealth that created the high life of the liners, just as surely in the end it was that which destroyed them.

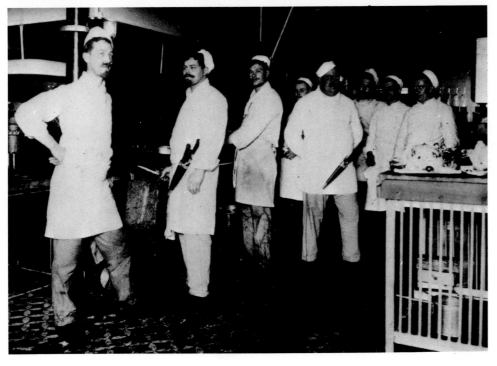

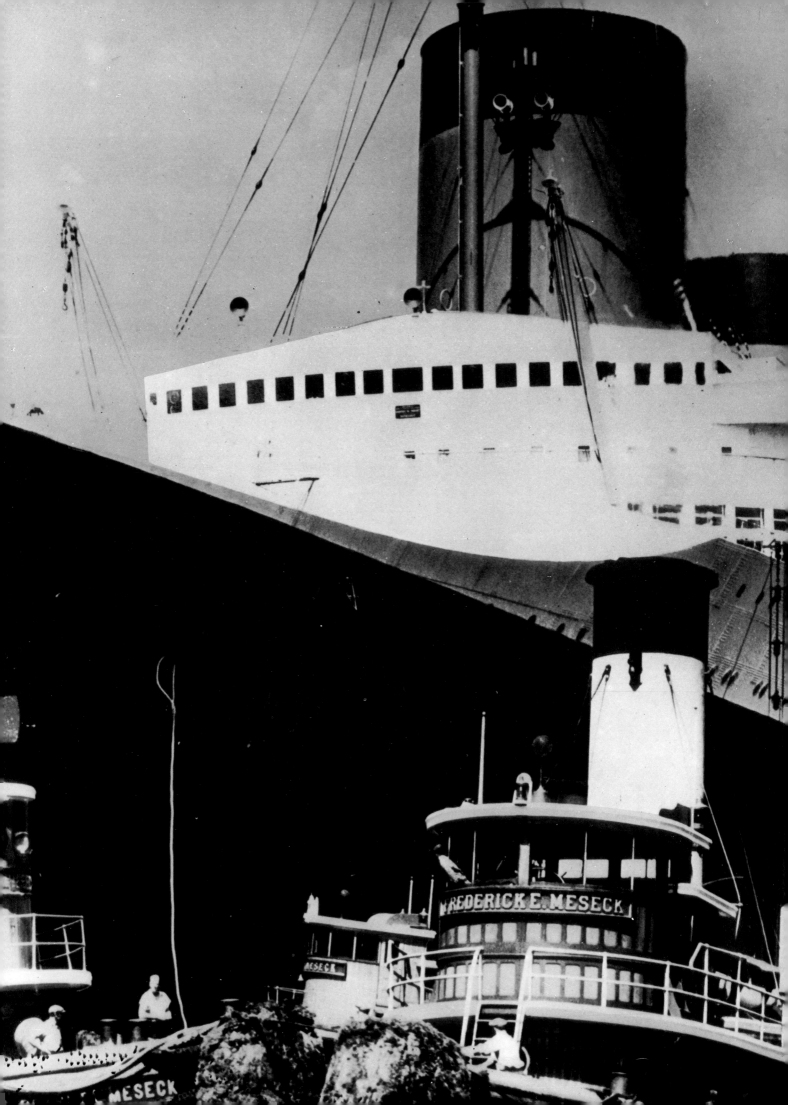

CHAPTER SIX

THE BLUE RIBAND

The Atlantic record-breakers

The Normandie
*after her record-
breaking voyage in
1935 : amidships she
carries a thirty-
meter Blue Riband.*

THE VERY NAME of the world's most glamorous speed record – the Blue Riband of the Atlantic – evokes exciting memories, although a quarter century has passed since the last holder of this remarkable trophy ploughed her triumphant way across the world's most dangerous ocean. The spectacle of the largest liners vying with each other for the fame and fortune, which came inevitably to the world's fastest ship, was one which stirred the imagination of people everywhere in the first half of this century.

The Blue Riband was an elusive trophy in many ways. The title originates from the blue insignia of the British Order of the Garter, the most exclusive decoration in the world. No one knows when the term was first used to describe the record for the fastest passage between Europe and America, but it was certainly in general use by the 1880s. Yet while the whole shipping world took an intense and compulsive interest in the record, in reality it never existed as a formal competition. There was no controlling committee, no entry forms, no official timekeepers, no specified course and some companies even chose to ignore the Blue Riband altogether, even after ensuring at considerable expense that their own ship was the new possessor of the title. As an example, when the *Queen Mary* crossed from the Bishop Rock to the Ambrose Light Vessel in August 1936 taking four days and twenty-seven minutes at an average speed of 30·14 knots, thus taking the record previously held by the *Normandie*, Cunard issued the following statement:

'We are only interested in having the liner officially designated the fastest. We deprecate record breaking voyages and we do not recognise the Blue Riband Trophy, which we shall not claim from the French liner *Normandie*.'

In contrast to this relatively hardheaded approach, the *Normandie* arrived in New York flying a 30-metre blue pennant (one for every knot of speed) when she took the record. The reluctance of some companies to publicize their intention to build a record-breaker is understandable, at least in commercial terms. The cost was considerable and meant a massive commitment of company funds. Nearly all the twentieth-century record-breakers also needed financial aid from national governments. With such sums involved, a failure to achieve the record would cause considerable embarrassment, both financial and in terms of national prestige. So all the while that company planners were scheming a new ship with the record in mind, the directors remained tight-lipped, attempting to impress on all that record-breaking was far from their thoughts. But when the record was won, then the cream of the Atlantic trade followed to fill the successful ship and repay her owner's investment in full.

All the big companies on the Atlantic flirted with the Blue Riband at some stage of their history, if only fleetingly. The Compagnie Générale Transatlantique maintained for years that comfort and luxury were preferable to speed, and eschewed record-breaking. Then, in a quick volte-face, it produced a superb expression of all three qualities in the *Normandie* of 1935. On the other hand, Hamburg-Amerika was so chastened by the capricious performance of its *Deutschland* of 1900 (she vibrated abominably) that the company deliberately ordered no further Blue Riband contenders.

But if the shipowners affected indifference to the Riband, the public's interest never flagged through all the 155 years that elapsed between the first voyage of the *Great Western* and that of the *United States*, last holder of the Atlantic crown. Even before the age of steam, when the American clippers were the undisputed Queens of the Western Ocean, the public on both sides of the Atlantic enthused over the achievements of *Flying Cloud*, *Columbia*, and *Sovereign of the Seas*, and argued closely the merits of each vessel. These sailing packets were no idlers. Given the right conditions (something that did not always happen) the sailing packets were fast. The *Andrew Jackson* once logged fourteen days only from Liverpool to New York. While the sailing packets could not be relied on to provide a scheduled service, there were enough of them to ensure a ship was always available to sail on time as announced, and they made fortunes for such men as Isaac Wright, owner of the Black Ball Line and his great rival of the Swallowtail Line, the New Bedford owner, Preserved Fish.

But the arrival of Sam Cunard's *Britannia* in 1840 meant the eventual doom of the sailing packets, and today the Liverpool paddler is usually credited as the first holder of the Blue Riband in the modern sense of the title.

The *Britannia* and her sisters were tight little ships and all varied slightly in dimensions. The *Britannia* herself was 1,156 tons and 207 feet in length. Inside her engine space, two single-cylinder side-lever engines were fitted, each with 72-inch diameter cylinders and an 82-inch stroke. In exchange for 38 tons of coal a day, these engines could provide up to 720 i.h.p. from four boilers, and turned the paddlewheels at a rate sufficient to produce 10 knots speed.

The first Cunard records were set up on the route between Liverpool and Halifax, Nova Scotia, and there were to be a number of start and finish points both east and west, as the years went by. Initially, the course was from Liverpool to Halifax, but when New York emerged as the main passenger port on the east coast, the western terminal point became Sandy Hook, the promontory at the harbor entrance. Later on, it was moved to the Ambrose Light Vessel. For most of the 19th century the eastern start line was Queenstown in Ireland but the North German Lloyd seldom called at this emigrant trading port, and when *Kaiser Wilhelm der Grosse* took the record in 1897 she started from the Needles, off England's Isle of Wight.

Later on, Cherbourg Breakwater, the Eddystone, Daunts Rock and the Bishop Rock Lighthouse all served their turn as marker posts, while the Italians made their starting mark Tarifa Point, fifteen

The flamboyant Hales Trophy for the Blue Riband donated in 1933 by a British Member of Parliament, Harold K. Hales. Surrounding the trophy are Blue Riband winners:
Arizona, 1879; Queen Mary, 1936; City of New York, 1889; Teutonic, 1889; Deutschland, 1900; Persia, 1856; United States, 1952.

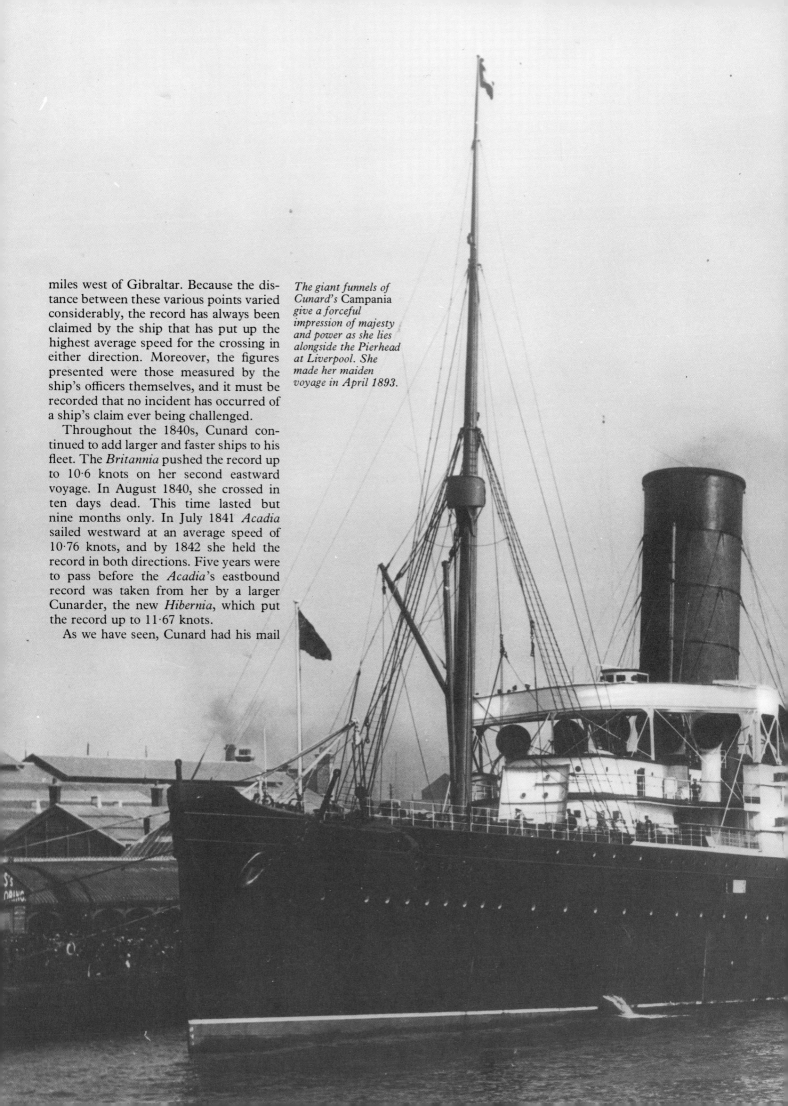

miles west of Gibraltar. Because the distance between these various points varied considerably, the record has always been claimed by the ship that has put up the highest average speed for the crossing in either direction. Moreover, the figures presented were those measured by the ship's officers themselves, and it must be recorded that no incident has occurred of a ship's claim ever being challenged.

Throughout the 1840s, Cunard continued to add larger and faster ships to his fleet. The *Britannia* pushed the record up to 10·6 knots on her second eastward voyage. In August 1840, she crossed in ten days dead. This time lasted but nine months only. In July 1841 *Acadia* sailed westward at an average speed of 10·76 knots, and by 1842 she held the record in both directions. Five years were to pass before the *Acadia*'s eastbound record was taken from her by a larger Cunarder, the new *Hibernia*, which put the record up to 11·67 knots.

As we have seen, Cunard had his mail

The giant funnels of Cunard's Campania *give a forceful impression of majesty and power as she lies alongside the Pierhead at Liverpool. She made her maiden voyage in April 1893.*

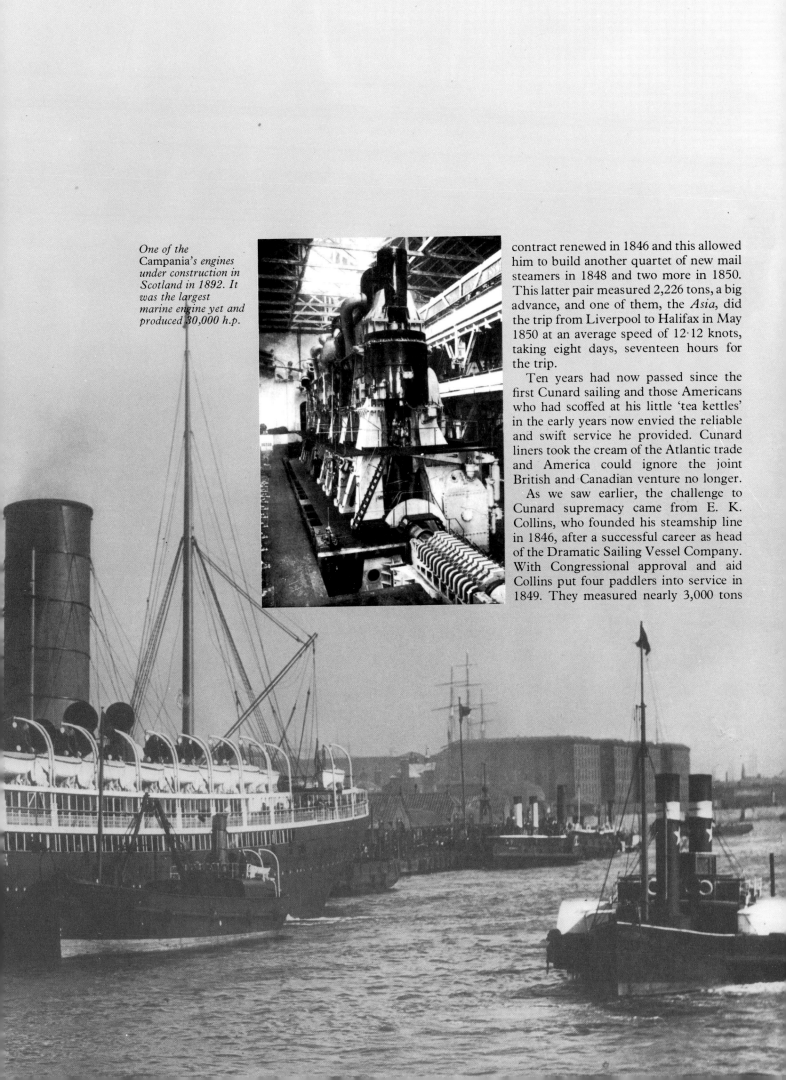

One of the Campania's engines under construction in Scotland in 1892. It was the largest marine engine yet and produced 30,000 h.p.

contract renewed in 1846 and this allowed him to build another quartet of new mail steamers in 1848 and two more in 1850. This latter pair measured 2,226 tons, a big advance, and one of them, the *Asia*, did the trip from Liverpool to Halifax in May 1850 at an average speed of 12·12 knots, taking eight days, seventeen hours for the trip.

Ten years had now passed since the first Cunard sailing and those Americans who had scoffed at his little 'tea kettles' in the early years now envied the reliable and swift service he provided. Cunard liners took the cream of the Atlantic trade and America could ignore the joint British and Canadian venture no longer.

As we saw earlier, the challenge to Cunard supremacy came from E. K. Collins, who founded his steamship line in 1846, after a successful career as head of the Dramatic Sailing Vessel Company. With Congressional approval and aid Collins put four paddlers into service in 1849. They measured nearly 3,000 tons

and contained much better accommodation than the Cunard ships. Their principal weakness lay in the lack of experience in America in the design and construction of large marine engines and several visits were paid to British engineering works. It was while returning home on the Cunard *Niagara* that Faron, consulting engineer to Collins, discovered that the Cunarder was working on a steam pressure of 13 p.s.i. when he fondly imagined her to be using not more than 10! The result of Faron's mid-ocean industrial espionage was a redesign of the engines for Collins' boats to incorporate monster side-lever machinery with 96-inch diameter cylinders stroking 120 inches and demanding 85 tons of coal per day to maintain 12 knots at 17 p.s.i. The cost of the ships shot up alarmingly, and Collins needed every dollar of his subsidy from Congress. Collins' first record holder was his *Pacific* which set up a westward record in 1851 of 13·01 knots, closely followed by *Baltic* with 13·17 knots.

The ship which took the record was the Cunard Line's ultimate paddle steamer, the *Scotia* of 1862. Cunard had continued with the well-tried paddle system much later than most companies. The screw was commonplace by 1850, and most economical in fuel consumption. But as the early two-bladed screws lacked much in efficiency, the paddle boats were faster, although they used large quantities of coal and made record holding an expensive affair.

Cunard was also late in the field with iron-built ships and the *Persia* of 1856 was his first. She crossed at 12·9 knots, not good enough to take the record, which finally fell in June 1862 when the beautiful *Scotia* went from Sandy Hook to Queenstown in eight days, twenty-two hours at a speed of 14·06 knots.

The *Scotia* was the last paddle-driven steamer to hold the Blue Riband. When she was built in 1862, Cunard ordered a screw steamer at the same time. The *China* at 2,529 tons was smaller than the *Scotia*, but she could carry 160 first-class and 771 steerage passengers at 13 knots, using 82 tons of coal every twenty-four hours. It took exactly twice that rate of coal consumption to keep the *Scotia* moving at 14 knots, and such figures dictated the demise of the paddler. The *China* became the first Cunard screw liner and she remained a popular ship on the Atlantic for sixteen years. After Cunard sold her in 1878, she remained active until 1906, when she vanished in the Pacific.

Cunard held onto the record until 1869. The *Scotia* was the last ship of Sir Samuel Cunard, who died in April 1865, aged 78. His great company was now one of the most important in Europe and it continued to follow the policies that Cunard had conceived.

In 1869, however, a new challenger entered the lists. This was the Inman Line, founded in Liverpool in 1850 and the ship was their *City of Brussels* of 3,747 tons. She crossed eastwards in seven days, twenty-two hours and three minutes, thus settling the arguments that had gone on since 1867, when both Cunard and Inman had claimed the prize with the *Russia* and the *City of Paris* respectively. It would take Cunard sixteen years to recover the record.

In 1872, the White Star Line went to Harland and Wolff at Belfast for the first time and the results of the deal were so good that Ismay's line never ordered a ship anywhere else for the remainder of its commercial existence. The *Adriatic* and *Baltic* both took the record in succession in 1872 and 1875. The *Baltic* was the first ship to make an average of 15 knots for the crossing, and both ships offered increased passenger comfort. Armchairs appeared in saloons for the first time and stewards could be summoned by the new-fangled electric bell!

Then the Inman Line hit back with *City of Berlin* in 1875 (15·41 knots) only to lose the record again to White Star in the following year when the Belfast yard produced another pair of crack liners, the *Germanic* and the *Britannic*. These sister ships held the record between them until 1879, the best speed being the *Britannic*'s

15·97 knots in December 1876. They were the first of the 'long ships', their designers deliberately giving them long sharp bows and clipper sterns for low hull resistance in the water and a rakish, speedy appearance. Yet, with characteristic Victorian prudence, they all retained four masts and cross-yards and were capable of proceeding on sail alone if they suffered engine failure and lost the power of their single screw. Fortunately such events were rare, and the single major accident of this type was the breakdown in mid-ocean of Inman's *City of Brussels* in February 1873. Wallowing along in mid-ocean minus her rudder and rudder post, she was found by her consort *City of Paris* and towed into Queenstown.

So reliable had engines become, that auxiliary sailing equipment was removed from most of the liners by the 1890s, but before then, the combination of raked masts, cross-yards, slim funnels and lean clipper-type hulls all made for an attractive appearance. The late Victorian liners are among the most beautiful ships ever built.

So the race went on. In the 1880s, the record changed hands on twelve occasions during that decade. The Liverpool-based Guion Line, American-owned but sailing under the British flag, produced the

Arizona in 1879 and followed her with the 6,400-ton *Alaska* in 1882, which became the first liner to cross in under seven days. No company made greater sacrifices for the sake of sheer speed than the Guion Line. So much space was given over to machinery that the resulting reduction of saleable passenger accommodation made both vessels unprofitable, and they both vibrated violently at high speed. Despite being thoroughly uncomfortable, the ships did well at first, and a third, the *Oregon*, was ordered. She took the record in April 1884, but lost it to the National liner *America* in June of the same year. Three months later, the *Oregon* took it back, but by then she was under the Cunard flag, as the cost of record-breaking had forced the Guion Line to withdraw from the race and sell its best ships.

The record was back with Cunard, held by the ship with which the company had expressly set out to beat with two new liners which they were about to take over from the builders. The famous Cunarders *Umbria* and *Etruria* of 1885 were excellent ships and together they took the record to just under 20 knots. True to Cunard's traditional conservatism, they relied on the well-tried single screw but were the last big liners to do so.

But they did possess steel hulls and electric lighting, innovations which had arrived in 1879.

Engineers had been experimenting with twin-screw propulsion for some time when Cunard built the *Umbria* and *Etruria*, and it was first used in Inman's reply to the Cunarders. This was the handsome *City of Paris* of 1889 and her sister *City of New York*, the first ships to top 10,000 tons and exceed 20 knots. The Inman ship took the record, but on 25 March 1890, the *City of Paris*, racing flat out to improve the margin over Cunard, fractured a crank-shaft and suffered severe engine-room damage. This setback checked Inman's plans and the title passed to the White Star's *Teutonic* in 1891. The *Teutonic* reduced the passage time down to five days,

sixteen hours and twenty-one minutes.

The *Teutonic* lost the record by 0·3 of a knot again to the *City of Paris* in 1892, but it was Inman's last throw. They lost it to Cunard in 1893, and that same year Tom Ismay decided that the future of the White Star lay in passenger comfort and cheaper passage tickets. The new Cunarders, the *Campania* and the *Lucunia* of 1893 put the record up to 22 knots and most people believed that figure to be the end of the line. To go faster meant a ship that was mostly engine, so much increased power being required to raise just an extra knot of speed. So people in the shipping world of 1893 were prepared to see Cunard hold the record for evermore.

They had reckoned, however, without Germany and her young, ambitious ruler, Kaiser Wilhelm II. German unity

in 1870 had provided the political edge to the powerful economic forces that emerged, as German industrial power grew in the eighties and nineties. Wilhelm envied the great colonial empires of Britain and France and he set out to win a similar position for his country. In a world where the steam engine was still the prime mover in transport, and which had yet to see the motor car or a radio set, the giant luxury liner was a colorful, sometimes awesome example of power, and Wilhelm was quick to grasp the implications of the national prestige attached to such ships.

With the Kaiser taking a personal interest, the North German Lloyd laid down two ships in 1895, and their very names, *Kaiser Wilhelm der Grosse* and *Kaiser Friedrich* served notice that Ger-

North German Lloyd's elegant Kaiser Wilhelm II turning in New York's Hudson River.

MAURETANIA
Luxury goes to sea

(QUADROPLE TURBINE) S.S. MAURETANIA. THE LARGEST VESSEL AFLOAT.

The introduction of the *Mauretania* and the *Aquitania* to the north Atlantic by Cunard in the first decade of the 20th century marked a decisive step forward in liner size and design. Regarded with great affection by many transatlantic passengers, the old 'Maury' also remained a 'rattler' to the end of her days.

A procession of early motor cars passes through the funnels.

Rank upon rank of furnaces grow in the boiler shop.

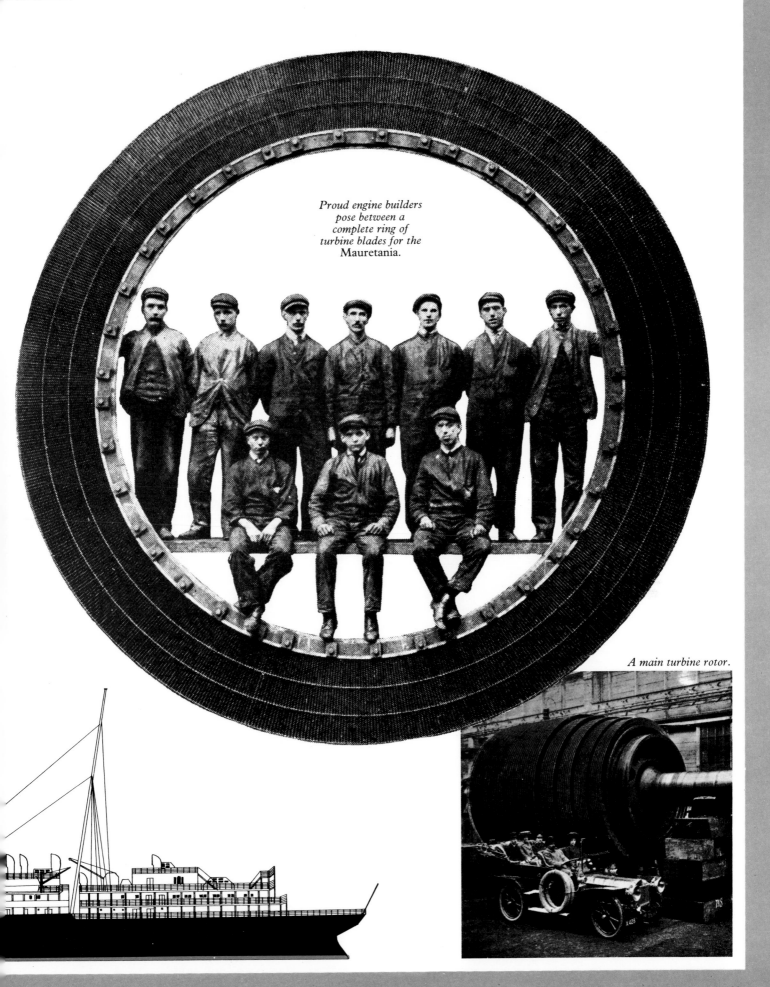

Proud engine builders pose between a complete ring of turbine blades for the Mauretania.

A main turbine rotor.

The Mauretania's *upper deck was a mass of ventilators supplying air for the hungry funaces far below.*

many was placing national prestige on the fortunes of these ships. The contract terms were tough. Trials were to include the maiden voyage and the ship could be returned to the yard if speed was not satisfactory. That is just what happened to the *Kaiser Friedrich*, which proved a failure.

The *Kaiser Wilhelm der Grosse* was launched in March 1897 in the presence of the Kaiser and 30,000 wildly cheering Germans. Wilhelm was to preside over a number of similar occasions in the twenty-one years remaining to him of his rule over Germany. The ship sailed on her maiden voyage on 19 September 1897 and in the following November took the record from Cunard at 22·35 knots, a margin of 0·7 knots, which she increased to a whole knot a year later.

British prestige had suffered a knock with this first German Blue Riband holder and the Germans followed her with four very similar ships in the next eight years. Each was a four-stacker like the *Kaiser Wilhelm der Grosse* and each a little larger and faster. All of them had huge reciprocating engines to get the necessary power, and at high speeds there was considerable vibration. The *Kaiser Wilhelm* lost the record in 1900 to Hapag's only flier, the *Deutschland*, also built by Vulkan at Stettin, which put the record

The Mauretania *ends her days in the breaker's yard at Rosyth at the end of 1935.*
Right : a trenchant American cartoon comment on Anglo-German rivalry for the Blue Riband after the Lusitania *took the record in 1907.*
Far right : the Mauretania *leaves Southampton for the last time on 1 July 1935, bound for the breaker's yard. Her topmasts have been cut down to allow her to pass under the Forth Bridge.*

190

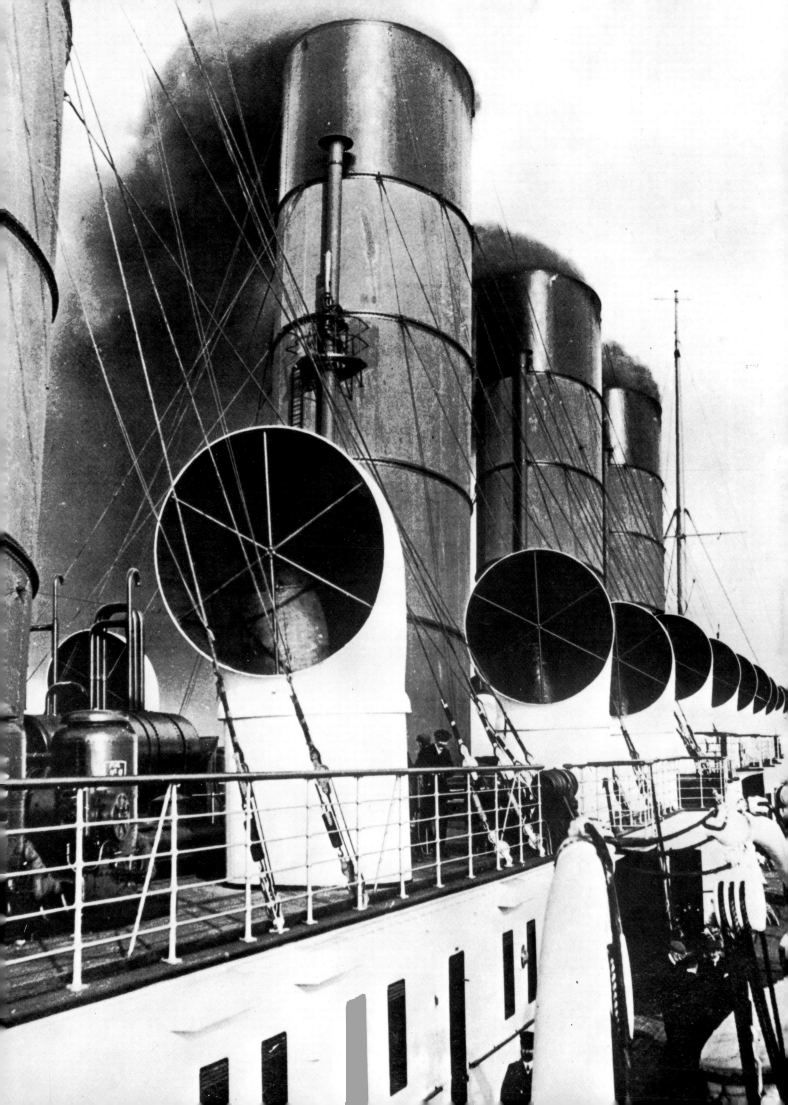

to 23·51 knots. But this latter ship was a sporadic performer and vibrated badly at any speed over 18 knots. Nevertheless, she and the two North German Lloyd ships that followed, the *Kronprinz Wilhelm* and the *Kaiser Wilhelm II* left Germany in firm control of the North Atlantic as the century turned.

In an age when every British child was taught that Britons really did rule the waves, it is not surprising that the British Government responded to this situation by giving Cunard a hefty subsidy to build the *Lusitania* and *Mauretania* of 1907. As we have seen, the challenge from Germany, the threat of Morgan's immense IMM corporation and Parsons' invention of the turbine all combined to produce the two wonderful liners that sailed from

the Clyde and the Tyne in the fall of 1907.

The *Lusitania* was first away. She sailed on trials in July and registered 26·45 knots over the measured mile off the Clyde coast at Skelmorlie. Somewhat sleeker in appearance than her sister, her 31,500 tons made her far and away the largest ship in the world and allowed her accommodation to be the best yet provided on the Atlantic.

The loss of the *Lusitania* in 1915, after only eight years in service, has led to her service record being ignored by historians who have concentrated on her terrible destruction. But she was every bit as fast as her famous sister, as she proved in 1909, when she crossed westward in four days, eleven hours and forty-two minutes at 25·85 knots.

If the word 'classic' can be applied to a ship, then the *Mauretania* of 1907 was such in every sense of the word. Large, fast and elegant, she queened it over every other liner on the Atlantic for a quarter of a century and retained her popularity with the public on both sides of that ocean through all those years. Her machinery performed with consistent reliability and possessed an amazing capacity for turning in improved performances when she was well past her twentieth year. Although forty years have passed since her final voyages, she still occupies a major place in any history of North Atlantic navigation.

The *Mauretania* took a while to settle down to her best performance and it was not until 1909 that she began putting in

Left : the stately
Europa *punches her
majestic way through
a north Atlantic gale.
Like her sister*
Bremen *she too held
the record in the early
thirties.
Below : the bulbous
bow, forerunner of
now standard naval
architectural practice,
stands out clearly in
this picture of the*
Europa *in dry dock.*

her best efforts. During her winter refit
at the end of 1908, she was given four-
bladed propellers to reduce severe vibration at high speeds. This was successful,
although the ship had the reputation of
being a 'rattler' all her days. More to the
point, the ship exceeded 27 knots with
her new screws and her speed on passage
went up. In October 1909 in severe
weather, she turned in a speed of 25·94
knots while rolling considerably and her
best crossing before the 1914 war was four
days, ten hours and fifty-one minutes for
an average speed of 26·06 knots.

There the matter was to rest for twenty
years. The *Mauretania* held the record
longer than any other ship and she steadily improved as the years went on. In
August 1924 she put the Riband speed up
to 26·25 knots, the improved performance being attributed to her conversion
to oil fuel, a transformation carried out
on all the big liners afloat in the early
twenties.

At last, when Cunard was considering
a replacement for the ageing express
liner, a challenger for her record appeared. The new ship was almost twice
the *Mauretania*'s size and power, and
came from Cunard's old enemies, the
North German Lloyd, in the shape of the

twin 52,000-ton superliners, the *Europa* and *Bremen*.

It had taken the German shipping industry just a decade to recover from the disasters of the First World War, and by the mid-twenties the North German Lloyd was ready to re-enter the Atlantic race from which the Cunarders had brusquely dismissed them over twenty years before. The *Europa* and the *Bremen* were big, squat vessels with ugly, short twin funnels which had to be lengthened later to prevent their depositing soot on the passenger decks. The ships were the first large liners to come down to two funnels (a distinct reduction from the *Mauretania*'s four!).

The new Germans were launched on succeeding days in August 1928 and the North German Lloyd planned *Europa* to be first away. But in March 1929, she was all but destroyed by fire in her builders' Hamburg yard. Flooded by an over-zealous fire brigade, the ship sank to rest on the dock bed, luckily in an upright position. The fire caused a year's delay and so it was the *Bremen* that sailed down the Weser on 16 July 1929 to challenge the old 'Maury'. She did all that her owners required of her and clipped eight hours off the Cunarder's best time.

In the weeks that followed, the Cunard board allowed Captain McNeal of the *Mauretania* to make one final riposte to the Germans. Chief Engineer Coleman and his team strained their valiant old engines to the limit. The *Bremen*'s record stood at 27·83 knots and when the *Mauretania* sped past the Bishop Rock at the end of her fastest passage ever, she had achieved an incredible 27·22 knots, just ·61 of a knot behind the brand new German. In her defeat she earned more fame than in any of her triumphs. In later years the old ship went cruising, smartened-up with a shining white hull, starting her last voyage from New York on the day the *Queen Mary* was launched, 26 September 1934. She was scrapped in 1935 and so passed the ship of which a great shiplover, President Franklin Roosevelt, wrote: 'Every ship has a soul, but the *Mauretania* had one you could talk to . . . as Captain Rostron once said to me, she had the manners and deportment of a great lady and behaved as such.'

In March of 1930, the *Europa* finally

Left : the giant twin sisters Bremen *and* Europa *rest quietly in their home port of Bremerhaven. Right : stylish North German Lloyd poster from the thirties. The* Columbus *was rebuilt to take on the appearance of the* Bremen *and the* Europa, *but she was out of her class alongside the two superliners.*

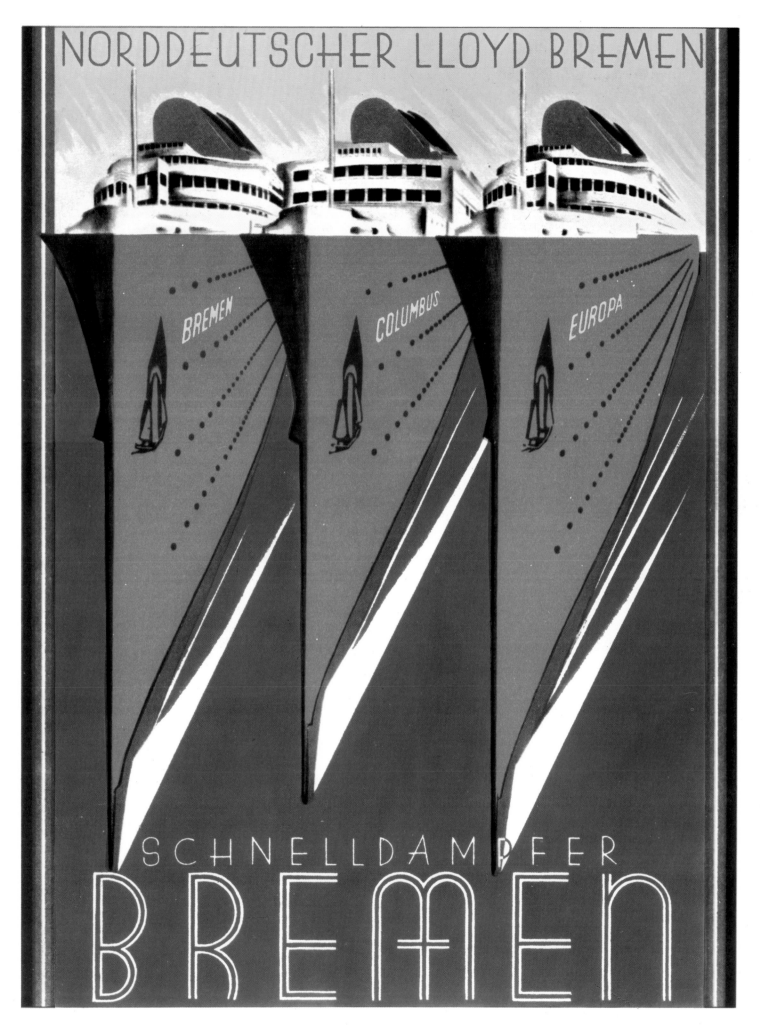

Italy's only Blue
Riband holder, the
51,000-ton Rex.
She took the record
away from the
Bremen in August
1933.

The huge gyro
stabilizers fitted in the
Italia liner Conte di
Savoia. The experi-
ment was only
partially successful.
Right: the popular
US Lines Manhattan
which, with her sister
Washington, took
over the American
services from the vast
and unprofitable
Leviathan.

went into service and took the westbound
record from the *Bremen*. But she was
never able to take the eastward prize and
by 1933, the *Bremen* held both again.

In 1930, the year of the *Europa*'s debut,
the great depression raged unchecked
through the economies of all western
industrial nations. One immediate outcome of the lean years (Atlantic traffic fell
off steeply) was that the big liner companies reluctantly sought salvation in
amalgamation, sometimes under hard
government pressure. Thus Hamburg-
Amerika and North German Lloyd passed
under the same control in 1931, although
they still traded independently. In Britain Cunard and White Star came together in the merger that allowed work
to go ahead on the first of the twin 80,000-
ton monsters, each with a 30-knot capability to maintain a weekly service.

In Italy also, considerations of national
prestige and economic commonsense
made Mussolini, the Italian dictator,
force the three top national companies
into the 'Italia' organization. Two of
the companies already had a big ship
on the stocks, both built with the idea
of cutting the time between Genoa and
New York to less than a week. The first
ship out was the *Rex* in 1932. She is
alleged to have taken her name from King
Victor Emmanuel, who was present at
her launch. Her owners are also said to
have suggested to Mussolini that the
second ship should honor the Duce by
assuming the title *Dux*. To everyone's
surprise, the dictator had an uncharacter-
istic lapse into modesty and refused, so
the ship became the *Conte di Savoia*.

The design of the *Rex* closely followed
that of the *Bremen*, but she had better
lines than the German and was a good-
looking ship. After a difficult maiden
voyage when she broke down off Gibraltar, the *Rex* took the record in August
1933 with a westbound crossing at 28·92
knots. Although the *Conte di Savoia* put
up some fast crossings, she never took
the record and is best remembered as
the ship which carried the first stabilizing
equipment fitted in a liner, three big

ITALIAN SUPER-LINER "REX" TODAY SMASHED ALL ATLANTIC SPEED RECORDS THEREBY CAPTURING BLUE RIBBON STOP TIME GIBRALTAR TO NEWYORK 4 DAYS 13 HOURS 58 MINUTES STOP AVERAGE SPEED 28.92 KNOTS STOP BEST DAYS RUN 736 MILES AT AVERAGE 29.61 KNOTS STOP=

ITALIAN LINF.

NORMANDIE
The finest liner ever built

Justly thought by many people to have the best claim to be considered the greatest triumph of passenger ship architecture, the *Normandie* was a happy combination of the best of the French engineering genius and flair for design and decoration. Externally, the soaring prow produced an impression of instant drama, while inside all the greatest artists of French interior design had been allowed to create vast rooms and spaces of stylish elegance. It was one of the saddest losses to the Atlantic route when this great ship did not survive the Second World War.

The ship's frame takes shape at the Penhoet yard at Saint Nazaire during 1931 and 1932.

Center : the rear pair of screws set abaft the huge balanced rudder. The four-bladed screws were replaced with a three-bladed pattern on April 1938 to cut down vibration.

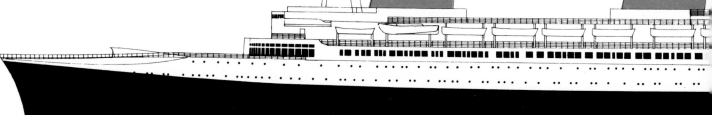

*Top: an alternator
for the turbo-electric
drive is slung on
board.
Above: a turbine
rotor.*

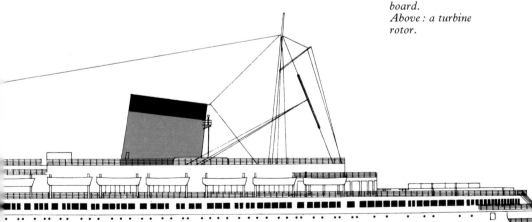

gyros whose sheer weight added more stability than their often sporadic performance. Passengers with tender stomachs would have to await the arrival of fin stabilizers after the 1939 war before their sufferings from sea sickness could be effectively eased.

The Depression years brought a new breed of ship to the North Atlantic run. Medium-sized, comfortable and with a speed that did not impose excessive demands for power, they could be operated at a profit without subsidy. The United States Lines' *Manhattan* was typical of these newcomers and their numbers grew during the thirties. Though few realized it, the age of the superliner was already in decline and only three other ships were ever to hold the Blue Riband.

Yet when they came, even as their heyday was ending, these ships expressed the apotheosis of the liner, the supreme achievement of the shipbuilder's art, and came near to fulfilling Brunel's dream of 100,000-ton vessels, so clearly prophesied a century before.

First into service was the French Line's massive *Normandie*, a ship years ahead of her time, with her streamlined hull, acres of uncluttered deck space, and turbo-electric engines. Like the *Rex* and the *Queen Mary*, she was built with government aid. Her creation owed much to national pride, and she fulfilled the expectations of every Frenchman when she took the Riband on her maiden voyage in 1935. The trip was mounted with masterly French élan. Heading the passenger list was Madame Lebrun, wife of

the French President, and the novelist Colette. For five hectic days, the centre of Parisian society was to be found 'somewhere at sea', many kilometres from the rue de la Paix!

Exactly a year later, Cunard White Star celebrated their new-found unity with the arrival of the *Queen Mary*, almost the size of the *Normandie*, and, in the event, a knot or two faster. The *Queen Mary*, however, had none of the French liner's style and elegance. Her cluttered decks and the layout of her public rooms made her an inconvenient ship in many ways, yet she remained a firm favorite with crew and public alike for thirty years. Through all that time, she was known to seamen as a 'happy' ship, a term hard to define, compounded of efficiency, satisfaction and content. Certainly the

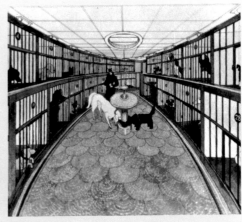

Top left : the winter garden on the promenade deck aft.
Top center : a first-class suite.
Above : the third funnel was a dummy and cosily contained the kennels for the dogs of the passengers.
Far left : a glimpse down the second funnel of the Normandie.
Left : the Normandie's divided funnel uptakes allowed her decorators to include long vistas through the centre of the ship. Here the Grand Staircase sweeps up to Baudry's statue, 'La Normandie'.

The Normandie is greeted in New York on 3 June 1935 by everything that can float (plus a blimp) at the end of her record maiden voyage.

Cunard triumphant! The Queens *burst out of a company poster from the fifties.*

Home from the wars, the Queen Mary *(right) ends her last voyage as a troopship, and joins the already refitted* Queen Elizabeth *at Southampton. At last the Cunard dream of a two-ship, weekly superservice can become a reality.*

food on the French liners was better, the *United States* was faster, but on the 'Mary' people were happy and got the services of the best crew on the Atlantic. To the British people as a whole, the 'Mary' represented more than anything their economic recovery and she generated considerable national pride. The term 'Queen Mary' became a synonym in English for anything of vast size and remains in common usage to this day. The *Queen Mary* took the Riband from the *Normandie* in August 1935, and for three years after that the pair pushed the record up to 31·20 knots, the *Normandie*'s best ever being in August 1937.

By now, a tangible award awaited the winner in the form of a metre-high, silver-gilt trophy, donated at his own expense by a British M.P., Harold K. Hales. Cunard would have none of this bauble, but both Transat and the United States Lines accepted it in their turn.

Instead, Cunard got on with the task of proving its staid *Queen* had the measure of her chic French rival. In August 1938, the issue was settled beyond doubt. The *Queen Mary* crossed from the Ambrose Light Vessel to Bishop Rock in three days, twenty hours and forty minutes, cutting one and a quarter hours from the *Normandie*'s time.

A year later, the world went to war again and the liners sailed off on seas more dangerous than at any time in history. When peace came, the *Bremen*, the *Rex*, the *Conte di Savoia* and the *Normandie* were no more and the *Europa* lay wrecked at Le Havre. The *Queen Mary*, joined at last by the *Queen Elizabeth*, was now able to offer the long promised weekly service to New York and Cunard exploited this supremacy in a brilliant 'Come to Britain' publicity campaign which earned the war-weakened British economy millions of dollars in hard cash. President Harry S. Truman estimated that the two *Queens* were earning for Britain at least $50 millions annually and his Government set out to restore the balance in favor of the United States.

The doyen of American naval architects, William Francis Gibbs, was commissioned to design the new vessel and Congress voted $48 millions of her eventual $75 millions cost. The keel was laid in February 1950, and, under a strict

*Looking every inch
an ocean greyhound,
the* United States
*pounds up the English
Channel at 35 knots,
the fastest liner ever
and the last holder of
the Blue Riband.*

security clampdown, Gibbs' designs began to take shape. The result was the *United States*, a gleaming fire-proof leviathan, whose light alloy upperworks allowed her dimensions to rival the *Queens* with a saving of 30,000 gross tons overall. Gibbs claimed that the only wood aboard was in the pianos and the butchers' blocks, omitting to mention that he had filled each bilge keel with balsa wood. Nevertheless his passion for safety led him to plead with the Steinway piano company for an aluminium grand piano (they refused) and he allowed no oil paintings in the ship's décor!

The *United States* was built in dry dock and floated out when nearly 90 per cent complete. Her two huge funnels, raked at an acute angle, rivalled those of the

Empress of Britain in size. Below, these were fed by only eight boilers, operating at nearly 1,000 p.s.i., and supplying four sets of Westinghouse turbines whose power has been kept secret to this day! A likely guess would be 240,000 s.h.p. at max. revs. Despite bad weather on her maiden voyage in July 1952 (a 60 m.p.h. gale was blowing as she passed Bishop Rock) the *United States* averaged an amazing 35·59 knots, cutting ten hours from the *Queen Mary*'s best time. She did almost as well on the return run to New York.

The Atlantic had a new champion and, as it proved, the last of the line. Even as she broke all previous records, and recovered the Riband last held for America by Collins' *Arctic* a century

before, the first jetliners were entering service. A plan for a sister ship did not proceed, and although it was seventeen years before the American taxpayer's pocket overcame his pride and the *United States* was withdrawn, she was obviously the last ship to be built with the Blue Riband in mind. Gibbs had died two years earlier, in September 1967. The *United States* was in New York on that day, and as she sailed past the designer's office in Lower Manhattan, three blasts erupted from the great ship's siren in tribute to the designer. It was as if a funeral note was sounded not only for Gibbs, but for Cunard, Ismay, Ballin and the rest, and for their great ships which were now passing into history just as their creators had done before them.

CHAPTER SEVEN

SHIPS OF THE LINE

Liners in the wars

The Union Castle's **Walmer Castle** *in the grotesque dazzle paint camouflage that was tried out on Allied warships and troopships during the First World War.*

209

HEN WAR BROKE OUT in 1914, only the actual time of its coming caused any surprise in Berlin or London. The event itself had been long expected and a good deal of preparation had been carried out on both sides.

Plans for such a war had affected the design and construction of both the British and German liner fleets. Germany had even formalized the position in the Naval Act of 1912, which called for all German merchant ships to carry guns below decks so that immediate mobilization could take place when war was declared. This was never fully carried out, but certainly the German Government planned to use some of its larger and faster ships in merchant raiding along the shipping routes of the world.

When the *Kaiser Wilhelm II* put into Southampton for repairs after a collision in the English Channel, for example, local dockyard workers were quick to notice that she had gun mountings fitted aboard. The British press took up the story instantly, but they were in no position to complain about German intentions. It was public knowledge that the conditions of the Admiralty loan to Cunard stipulated that the *Mauretania* and the *Lusitania* would become armed merchant cruisers in time of war.

It seems strange that naval planners believed so firmly that a large, unarmored passenger ship could be converted into a man-of-war. Yet the idea persisted in both the British and German Admiralties. In the event, the experience gained in the early months of the 1914 war proved conclusively that these large vessels, using prodigious quantities of coal which had to be supplied to outlandish locations anywhere on the globe, were entirely unsuitable for use as armed merchant cruisers. To their credit, the German admirals learned the lesson and in the Second World War operated small, fast converted freighters for raiding, of which Kapitän Rogge's *Atlantis* was a brilliant example. The British continued to use liners as cruisers as late as 1942 and suffered inevitable casualties.

The moment hostilities commenced in 1914, von Tirpitz at the German Admiralty wirelessed immediate war instructions to all German ships at sea. Those who had a chance of getting home to Germany were ordered to do so. Most of the remainder were told to put into neutral ports, while a few others were ordered away to little known anchorages, there to begin preparation for more sinister business.

Hapag's giant *Vaterland* was tied up in New York, having just finished her fourth Atlantic crossing westwards. Commodore Hans Ruser was ordered to lay up his ship rather than risk her running the British blockade. She lay at her Hoboken berth, attended only by a skeleton crew, for two and a half years until America entered the war on 4 April 1917. Then the *Vaterland*, together with several other big German liners, was taken over by the United States Navy. The obvious use for this windfall of war was trooping. Included with the *Vaterland* in this 'Fleet the Kaiser built us', as War Secretary Josephus Daniels gleefully called it, were *Kaiser Wilhelm II*, *Kronprinz Wilhelm*, *Kronprinzessin Cecilie*, *Amerika* and *George Washington*. In the next two years, they made up the bulk of the fleet which, together with the British liners, carried over two million Americans to the Euro-

pean war zone and then brought them home after the Armistice.

The *Vaterland* herself was renamed *Leviathan*, some say at the suggestion of President Woodrow Wilson and his wife. To crew and troops alike she became the 'Levi Nathan'. Of the two remaining giants owned by Hapag, the *Imperator* was safe in Hamburg and the *Bismarck* was incomplete. Neither ship took part in the war, and when the fighting was done the *Imperator* was sent to sea as yet another troopship for the Americans. It is said that they were thoroughly dissatisfied with her performance and were relieved when she was allocated to the British at the Peace Conference to become the Cunard flagship *Berengaria*. In similar fashion, *Bismarck* was completed in Germany and then, after her trials under, ironically, the great Hans Ruser, she was handed over to the White Star Line in March 1922. The line gave her one of the great traditional names of their fleet – *Majestic* – a happy choice for the ship that would for the next fifteen years be the 'World's Largest'.

As the news of war came, the urgent need for those ships at sea became a rapid dash for safety. Both the *Olympic* and the *Mauretania* were outward bound for New York. The White Star flagship held her course, upped speed to the maximum her boilers would take and made it safely to the Ambrose Light. The *Mauretania*, on the other hand, closed down her radio, blacked out the ship, and with a double shift of stokers, made a dash for Halifax, Nova Scotia. She got there safely, achiev-

ing a spectacular 28 knots in the process! Other ships were destined to make longer detours before a safe haven was reached. One of these was North German Lloyd's *Kronprinzessin Cecilie*. Commanded by one of the Atlantic's most popular skippers, a tough moustachioed salt called Karl Polack, the *Cecilie* was four days out from New York for Bremen with 1,200 passengers and $12 millions in gold and silver bullion in her strong room.

She was an obvious prize for any patrolling Allied cruiser. In the event, Polack was advised by radio that two French warships were on his tail and he acted characteristically. Without announcing his intentions to his passengers, he put the ship about and headed back for the American coast. The *Kronprinzessin Cecilie* possessed the largest reciprocating engines ever fitted in a ship and, flat out, these two jumping giants reduced the passenger accommodation to a hell of vibration. As the ship rattled her way back, Polack's crew painted the tops of the yellow funnels black in the hope that she might pass for the White Star's *Olympic*.

These were the days when the first-class passenger list of most of the big liners could muster a handful of millionaires and Polack was soon faced by a grim-faced delegation of American tycoons who solemnly offered to buy the ship from him so that he could run the Stars and Stripes, so defying any Allied cruiser to harm the ship! Stonily, Polack refused, and after running at full speed through a fog (to the horror of his terrified passengers!) he turned up in the remote

Bar Harbor in Maine where superliners were not considered regular callers. On arrival, he found he and his ship were international celebrities, as there had been worldwide speculation over the whereabouts of the 'Missing Treasure Ship'.

But other ships went to war. In the beginning it was a gentleman's war, played out with a gallant but unrealistic chivalry on both sides. While the Germans were hastily converting their available liners at remote anchorages, the British had carried on refitting liners as merchant cruisers, in some cases in as little as forty-eight hours. This process consisted of shipping eight 4·7 inch guns and stocks of ammunition, a coat of grey paint and a change of crew. Then the liners went to war. Fortunately, an attempt to use the big Cunarders as cruisers came to nothing. The *Aquitania* and *Mauretania* both received their armament, but a collision on her first day of service made the *Aquitania* put back for repairs, and experience of the *Mauretania*'s great appetite for coal convinced the Admiralty that the whole exercise was not economic. So both big Cunarders were sent trooping and later used as hospital ships, both roles in which they excelled. The *Lusitania* remained on the passenger run between Liverpool and New York.

As events turned out, only five of the forty-two German liners at sea when war broke actually became raiders. The remainder were blockaded in neutral ports. But of the five that fought, two were famous names indeed. Once the fastest

Right : a shell hole in the Carmania made by a German 4 inch gun. There were over 300 similar holes in the Cunarder. Far right : the Carmania's bridge after the action. Below : one of Carmania's 4·7 inch guns.

Opposite page : the protagonists, left. The Hamburg Sud-Amerika's Cap Trafalgar ; in 1914 she was virtually a new ship. The third funnel was a dummy and was removed when the ship was fitted out as a raider. Right, the Cunard liner Carmania on her trials in 1903. She was the first turbine-driven steamer owned by Cunard.

ship afloat when she won the Blue Riband in 1897, the *Kaiser Wilhelm der Grosse* was still good for 22 knots. Painted black overall and fitted out with six 4 inch guns, the old ship sailed from Bremerhaven on 4 August 1914 under Kapitän Reymann. Going north about past Iceland, she evaded British patrols, sank a trawler, the *Tubal Cain*, hardly a worthwhile prize at 250 tons, and then made for the Canaries. By this time very short of coal, Reymann had a stroke of luck when he captured the Union Castle liner *Galician* (5,000 tons) and three other ships, all in two days. Having released two of his prizes because of the presence of women and children on board (a gallant act that was to let the British know his whereabouts), Reymann steamed to the Spanish colony of Rio de Oro on the coast of the Sahara. Here, despite the protests of the local Spanish officials, he proceeded to coal ship for an incredible nine days stay and he was still coaling when, on 26 August, the British cruiser H.M.S. *Highflyer* arrived on the scene. Refusing Captain Buller's ultimatum to surrender, the old record holder went into action against impossible odds, since the *Highflyer* carried eleven 6 inch guns. A hero to the last, Reymann sent 400 of his men away in boats and fought the action with a skeleton crew. The result was inevitable. After an hour, the ship that was once the pride of the Atlantic was a sinking ruin. She had been too big and needed too much coal to fight a war.

The month following the sinking of the *Kaiser Wilhelm der Grosse*, two other liners slogged it out in one of the most memorable single-ship actions of either of the World Wars. The Cunard liner *Carmania* (19,500 tons) had been converted at Liverpool at the outbreak of war and now carried eight 4·7 inch guns. On 14 September 1914 she was approaching the tiny Brazilian island of Trinidada on a routine search. At dawn on that day, she came on the Hamburg Sud-Amerika liner *Cap Trafalgar* (18,500 tons), which was coaling at one of the island anchorages. The *Cap Trafalgar* had been armed at sea when she had received two 4 inch guns and six machine-guns from the German gunboat *Eber* at the mouth of the Plate at the war's onset. The gunboat's skipper, Kapitän-leutnant Wirth, took command and the ship sailed northwards as a commerce raider. She could produce 17·5 knots to the *Carmania*'s 16, but the British had a definite advantage in weight of armament. Wirth made a run for it with the Cunarder in hot pursuit, and the action opened at 8,000 yards range. Wirth then ran up the Imperial Ensign and turned to give battle.

Wirth's only hope lay in closing the range and so allowing his heavy machine-guns to rake the *Carmania*'s open decks and make her gun positions untenable. He therefore closed the range to 3,000 yards and was for a time almost within sight of his objective. The *Cap Trafalgar* poured a heavy stream of lead into the British ship and in all hit her with seventy-nine shells as well. Heavy fires broke out on the *Carmania*, but Captain Grant held on and concentrated his fire on the German ship's waterline. This tactic paid off, as shell after shell plunged deep inside the blazing *Cap Trafalgar*.

The irony of the action was that, had he but known it, Wirth's guns outranged those of his British opponent by 2,000 yards, but now Wirth lay dead and his ship was dying with him. The *Cap Trafalgar* keeled over and sank, taking most of her crew with her and leaving the crew of *Carmania* the daunting task of saving their shattered vessel from fires started during the action. The battle off Trinidada proved once again that liner hulls could not stand up to even the smallest guns mounted by the world's navies and the firing line was no place for them.

Most successful by far of the German liners that became raiders was the former North German Lloyd Blue Riband holder *Kronprinz Wilhelm* (15,000 tons). She was the only large German ship to slip past the British cruisers outside New York at the outbreak of war and she sped south, carrying 2,000 tons more coal than

The action between Carmania *and* Cap Trafalgar *at its height.*

The warning to travellers issued by the German embassy is prominently displayed below Cunard's Sailing Notice in a New York newspaper. Saturday, 1 May proved to be Lusitania's *final sailing date!*

Exhausted survivors struggle away from the side of the sinking Lusitania. *Allied war artists and writers made much anti-German propaganda from the incident, which led eventually to America's entry into the war.*

CUNARD

EUROPE VIA LIVERPOOL

LUSITANIA

Fastest and Largest Steamer
now in Atlantic Service Sails
SATURDAY, MAY 1, 10 A.M.
Transylvania. . . Fri., May 7, 5 P.M.
Orduna. . Tues., May 18, 10 A.M.
Tuscania. . . . Fri., May 21, 5 P.M.
LUSITANIA, Sat., May 29, 10 A.M.
Transylvania. . . Fri., June 4, 5 P.M.

Gibraltar - Genoa - Naples - Piraeus
S.S. Carpathia, Thur., May 13, NOON

ROUND THE WORLD TOURS
Through bookings to all principal Ports
of the World.
COMPANY'S OFFICE 21-24 STATE ST., N.Y.

NOTICE!

TRAVELLERS intending to embark on the Atlantic voyage are reminded that a state of war exists between Germany and her allies and Great Britain and her allies; that the zone of war includes the waters adjacent to the British Isles; that, in accordance with formal notice given by the Imperial German Government, vessels flying the flag of Great Britain, or of any of her allies, are liable to destruction in those waters and that travellers sailing in the war zone on ships of Great Britain or her allies do so at their own risk.

IMPERIAL GERMAN EMBASSY.

WASHINGTON, D. C., APRIL 22, 1915.

she needed to reach Europe. In the West Indies she rendezvoused with the German cruiser *Karlsruhe* and took aboard two 81mm guns and rifles and ammunition. In the process of doing so, the Germans were interrupted by the British cruiser H.M.S. *Suffolk* and had to make a run for it. After that, the *Kronprinz Wilhelm* operated in the Atlantic, while German naval attachés on both sides of the ocean sweated over the logistical problem of keeping the big liner supplied with 500 tons of coal a day. In this they were successful, for the ship's cruise took her 20,000 miles and she consumed 20,000 tons of coal in the process. Operating over 37,000 square miles of the Atlantic, she destroyed twenty-six Allied freighters, and when at last lack of fuel and supplies forced her to abandon her mission, Kapitän Thierfelder avoided the patrols sent to look for him and took his big old raider safely into Chesapeake Bay on 10 April 1915, after eight months at sea.

The return of the *Kronprinz Wilhelm* to New York marked the end of the German Navy's attempt to interrupt Allied shipping by surface raiding. In the meantime the Royal Navy had imposed a tight blockade on Germany and, for the determined von Tirpitz, there remained only one alternative weapon – Germany's growing fleet of submarines, which now included some long range boats which could remain at sea away from their bases for several weeks. The accepted rules of warfare in 1915 demanded that a submarine captain should surface and warn his intended victim before taking any aggressive action, a sure way of betraying his position to any enemy warship that was in the vicinity. It was obvious that, as the war developed, this convention was bound to be ignored, and that while military action against non-combatants was unthinkable in 1915, the world would become all too familiar with it in the quarter century that was to follow.

As already mentioned, the Cunard *Lusitania* had been retained on the north Atlantic mail run, after the Admiralty had decided against her use as a merchant

cruiser. This may have been due to the experience with the *Aquitania* and *Mauretania*, but whatever the reason, Cunard was allowed to sail her once a month on her old service, with six of her boilers shut down and a maximum available speed of 21 knots. This was considered sufficient to keep her out of trouble, but as further safety precautions, all watertight doors were closed during the voyage and in British home waters, all boats were swung out and lookouts doubled.

Two days after the German warning in the American papers, the *Lusitania*, under command of Captain William Turner, one of Cunard's senior masters,

left New York with 1,159 passengers and, allegedly, a small quantity of munitions. By Friday 7 May, she had reached the Irish coast and also received Admiralty warnings that U-boats were to be expected in the area of the Fastnet. At 1410 hours, in clear and warm weather, *Lusitania* was torpedoed by U.20 (Kapitänleutnant Walter Schwieger) with a single torpedo in the starboard side between the first and second funnels. The ship took an immediate list to starboard, making it impossible to launch any of her portside boats; at the same time she drove on, unable to stop as the explosion had destroyed her engine-room controls. In twenty minutes it was all over. *Lusitania* went down 10 miles south of the Old Head of Kinsale with the loss of 1,198 lives including 124 United States citizens. Will Turner went down with her but was plucked from the water by one of the armada of small craft which hurried to answer the *Lusitania*'s distress signals. He survived to answer awkward Court of Inquiry questions about his failure to carry out Admiralty instructions to zig-zag in British waters after U-boat reports.

Immediately, international controversy flared. Britain accused Germany of violation of International Law and associated war crimes. Germany countered with the claim that the *Lusitania* was carrying war material, an accusation that modern research in American naval archives has gone far to substantiate. Two facts that seem clear are that Schwieger fired but one torpedo, although all survivors swore that two explosions rocked the ship. That there were two explosions is undeniable on the strength of the evidence, but the cause of the second remains controversial to this day – either it was a boiler explosion or the clandestine cargo of explosives! Whatever the truth of the matter, a wave of public indignation swept Britain and America and anti-German sentiment created by the incident did much to prepare American public opinion for the Declaration of War by President Wilson in 1917.

Following their earlier failure as mer-

chant cruisers, the big British Atlantic liners were put into active use again during the unhappy Gallipoli campaign which started in mid-1915. This time the big ships were used for a purpose for which they had been designed – the transport over long distances of large numbers of people, and they proved very good at it. Thousands of troops were moved into the Aegean port of Mudros by the *Aquitania*, the *Mauretania*, the *Olympic* and the French Line's *France*. Later in the campaign, as the toll of wounded mounted, several of the big liners became hospital ships and these included the largest British ship at that time, the *Britannic* (48,158 tons), larger sister of the *Olympic* and the *Titanic*, which was incomplete at Belfast at the outbreak of war and had been fitted out, not for the luxury service to New York, but for a more spartan service to wounded troops.

She was ready by December 1915 and carried out five round trips to the Mediterranean in the spring and early summer of 1916. Then, on 1 November of that year, outward bound from Naples to Mudros, she struck a mine in the Zea Channel four miles west of Port St. Nikolo. Providentially there were no wounded on board, although over 1,100 medical staff and crew were manning the ship. The evacuation was orderly, weather conditions perfect and fatal casualties were limited to twenty-eight as a result. But the White Star Line had lost the second of the great trio of ships planned with such care in the pre-war years.

By the beginning of 1916, over twelve million tons of Allied shipping were serving as transports, auxiliary cruisers or supply ships and severe losses were inflicted by the U-boats. In the first two years of the war, Britain lost 338 ocean-going and 264 coastal vessels to enemy action and the total increased alarmingly in 1916 and 1917, when the country was brought to the point of starvation by the submarine campaign. But the use of the convoy system and the building of small fast escort vessels led to the eventual defeat of the U-boats, which suffered

The New York World *on the day after the sinking. Already reports speak of two torpedoes, although Schwieger always claimed that he only fired one.*

SETTING THE PACE!

91,400 WORLD AHS LAST JANUARY 37,169 More Than the Herald.
98,748 WORLD AHS LAST FEBRUARY 38,557 More Than the Herald.
107,781 WORLD AHS LAST MARCH 52,269 More Than the Herald.
119,673 WORLD AHS LAST MONTH 52,913 More Than the Herald.

Advertise in the Big SUNDAY WORLD To-Morrow!

"Circulation Books Open to All."

The World.

"Circulation Books Open to All."

VOL. LV. NO. 19,618.

NEW YORK, SATURDAY, MAY 8, 1915.

PRICE ONE CENT in Greater New York and Jersey City. TWO CENTS outside of Greater New York, Jersey City and on trains.

WEATHER FORECAST—Fair to-day and to-morrow; fresh to strong southwest wind is not wind.

DON'T FORGET!

Remarkable Photos of War from Men Who Man Them & Will Try the NAVAL REVIEW Edition of the Light Page Pictorial Supplement of The Sunday World To-Morrow!

TWO TORPEDOES SINK LUSITANIA; MANY AMERICANS AMONG 1,200 LOST; PRESIDENT, STUNNED, IN SECLUSION.

LUSITANIA, HER CAPTAIN, AND PLACE WHERE SHE WAS HIT

S.S. LUSITANIA

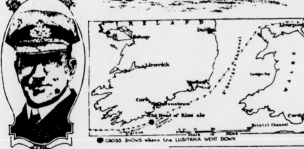

CROSS SHOWS where the LUSITANIA WENT DOWN

Captain W. T. TURNER

Liner Attacked Supposedly by German Submarine Off the Irish Coast, and Goes Down in Fifteen Minutes—Luncheon Being Served at the Time—Survivors Picked Up From Lifeboats and Taken to Queenstown, Forty Miles Distant—Regarding 1,254 Passengers and 850 of Crew Aboard, Cunard Line Says: "First Officer Jones Thinks 500 to 600 Are Saved"—Ship Left New York Last Saturday With Many Americans, Including Prominent New Yorkers, Who Disregarded German Warning Not to Sail.

(Special Cable Despatch to The World.)

LONDON, May 8.—The Cunard liner Lusitania was torpedoed, supposedly by German submarines, shortly after 2 o'clock yesterday afternoon, ten miles off the Old Head of Kinsale, on the south coast of Munster, Ireland.

She sank fifteen minutes later. The company states that no warning was given her.

Passengers and crew, the Lusitania carried 2,104 persons when she sailed from New York, on May 1. The meagre, confused reports so far received make it uncertain how many of these have been saved.

A steward of the first boat that reached Queenstown—forty miles by sea from Kinsale—with survivors from the liner, said he feared that 900 lives had been lost.

This despatch came from Queenstown at 1.10 A. M.:

"The tug Stormcock has returned here, bringing about 150 survivors of the Lusitania, principally passengers, among whom were many women, several of the crew and one steward.

"Describing the experience of the Lusitania the steward said:

" 'The passengers were at lunch when a submarine came up and fired two torpedoes, which struck the Lusitania on the starboard side, one forward and the other in the engineroom. They caused terrific explosions.

" 'Capt. Turner immediately ordered the boats out. The ship began to list badly immediately.

" 'Ten boats were put into the water, and between 400 and 500 passengers entered them. The boat in which I was approached the land with three other boats, and we were picked up shortly after 4 o'clock by the Stormcock.

" 'I fear that few of the officers were saved. They acted bravely.

" 'There was only fifteen minutes from the time the ship was struck until she foundered, going down bow foremost. It was a dreadful sight.'

"Two other steamers with survivors are approaching Queenstown."

An official statement issued by the Cunard Steamship Company said:

"First Officer Jones thinks from 500 to 600 were saved. This includes passengers and crew, and is only estimated."

A despatch to the Chronicle from Queenstown says that "seven torpedoes were discharged from the German attacking craft, one of them striking the Lusitania amidships."

This would indicate that at least two submarines were arrayed against the liner. Even the newest type of the undersea boats carries but six tubes, and most of them have only four.

A despatch coming from Kinsale at 7 o'clock says that at 3.30 two lifeboats were intercepted six miles off Old Head by the motorboat Elizabeth and convoyed by a Cork tug, which took from one 63 passengers and from the other 16, most of them women and children. They were taken to Queenstown instead of to Kinsale, whither they were bound.

These survivors said that the Lusitania got two torpedoes, the first of which struck her on the port side. She canted toward the land, and received the second on the starboard side.

They said a heavy list to port followed, and the Lusitania remained afloat for only ten minutes, and only six lifeboats could be launched. These contained about 300 passengers.

Other reports say that the first of the torpedoes struck the liner near her bows, the second tearing its way into her engine room. Terrific explosions followed, and great volumes of water poured in through the rents.

A midnight report to the Chronicle says that the number of survivors at Queenstown is 520.

An Admiralty report said that between 500 and 600 survivors have been landed at Queenstown, "including many hospital cases, some of whom have died."

In this report it is added that some were also landed at Kinsale, "the number not having yet been received." Private telegrams say that "several hundred passengers" have been landed at Clonakilty, not far from Kinsale.

WASHINGTON, SILENT, AWAITS ADVICES ON AMERICANS' FATE

Wilson, After Receiving Official News From Queenstown Saying "Probably Many Survivors; Rescue Work Progressing Favorably," Leaves White House in Drizzle—Calls No Counsellors or Conferrees on His Return—Situation Most Tense Since Spanish-American War—"Strict Accountability Note" to Germany Is Recalled.

(Special to The World.)

WASHINGTON, May 7.—The White House and State Department at 10 to-night received this message from Wesley Frost, the American Consul at Queenstown, Ireland:

"Lusitania sunk at 2.30. Probably many survivors. Rescue work progressing favorably."

The message also asked the State Department if Consul Frost should cable a list of the American survivors, to which the department replied that such a list should be sent immediately.

President Wilson had just finished dining when the message was presented to him. It appeared to stun him, because earlier messages had indicated that as passengers had been lost, and therefore America could not be vitally involved.

A few minutes later two Secret Service men, who were on guard at the offices, were surprised when a puffing policeman, who guards the cabin door at the White House, rushed in and told them that the President had just left the house unaccompanied.

As the Secret Service men rushed through the parking they saw the President cross Pennsylvania Avenue and take a course due north through National Street. He appeared to be oblivious to the light drizzle which was falling, and to the passersby who scampered through the streets.

"Extra, extra! many lives lost on the Lusitania! Americans among the dead!"

The President appeared to be deep in thought as he walked through

(Continued on Fourth)

CUNARD OFFICES CLOSE AS LIST OF THE DEAD GROWS

Action Unexplained, but an Official Said at 11 o'Clock Only 509 Had Been Accounted For, and Four of These Were Dead—Hoped to Hear of 200 More by Morning.

The New York offices of the Cunard Line, after giving out a series of bulletins during the early evening, were unexpectedly closed shortly after 11 o'clock last night. All the line officials, however, remained on duty, although reporters and other outsiders were shut out. No explanation was given for this action, nor would officials reply to the question whether their sudden decision was due to either even worse news than given out, or to a hint from the British Government that the censorship would interpose.

"We hope for a couple of hundred more in the morning," he said.

Eight persons so far are known to the Cunard officials by name as having been saved. They are Mrs. Bretherton of Los Angeles, Cal., and her two children, one and two years old; Mrs. H. B. Lassetter of Sydney, Australia, and her son, F. Lassetter; George A. Kessler of New York, Miss Jessie Taft Smith of Braceville, O., and Miss Irene Paynter of Los Angeles, Cal.

3 Years' Work—Cost, $8,000,000; Sunk by Torpedo—Cost, $4,000

The following estimate may serve to present to the readers of The World a mental picture of the giant liner Lusitania in comparison with the small but deadly submarine torpedo that destroyed her:

Average length of a torpedo	16 feet
Length of the Lusitania	790 feet
Average cost of a torpedo	$4,000
Approximate cost of the Lusitania	$8,000,000
Time required to make and test a torpedo	3 months
Time required to build the Lusitania	3 years
AND
TIME REQUIRED TO DESTROY THE LUSITANIA WITH A TORPEDO	15 minutes

Hospitals at Queenstown Caring for Lusitania's Survivors; Dead and Wounded Arriving on Boats From Scene of Disaster

Cork newspapers say that the number taken to Clonakilty was 300. Official figures make the number at Kinsale eleven.

The only figures so far offered by Queenstown account for 506 survivors and 4 dead. Of these the tug Stormcock was reported to have brought in about 100 passengers and crew; the trawlers Bock and Indian Empire have about 700, the tug Flying Fish about 100 and three torpedo boats 44 and 4 dead.

In making these figures known, the Admiralty announced that it was not withholding any terrible news, but that it would decline to pass despatches based merely on rumor.

The naval and military hospitals at Queenstown are receiving the wounded as they arrive. The survivors are being cared for at hotels and boarding houses, and because "their immediate wants must be given out first consideration," neither the Cunard officials nor the Admiralty will attempt to make up any list of them to-night, it is announced.

The first word that Queenstown had of the disaster was when she picked up this wireless call from the liner:

"Want assistance. Listing badly."

What this meant needed no deduction. During this week alone twenty-eight vessels had been sunk or damaged in the war zone that Germany had established about the British Isles.

Admiral Oaks lost not a moment, therefore, in despatching to Kinsale every available tug and steam trawler. The tugs Warrior, Stormcock and Julia led the procession, with five trawlers and a tug-towed lifeboat in their wake. Steamers in the vicinity picked up the liner's calls and started to her aid.

Sank Quickly After Being Struck.

From the signal station at the Old Head of Kinsale, the Lusitania was seen at 2.12 o'clock to be in distress. At 2.33 she had completely disappeared, according to reports to the Admiralty.

It was officially announced last night that she remained afloat "at least twenty minutes after being torpedoed." At that time, it was added, "twenty boats were on the spot."

How many of these were the Lusitania's own lifeboats there is yet no way of determining. It has been stated, however, that she had time to clear most, if not all, of them away. She carried boats and rafts enough for at least 300 more persons than she had in her company.

At midnight the Admiralty was still without information as to the heavy losses in the later years of the war.

One of them was particularly unlucky in a spectacular way. On 12 May 1918, U.103 was lying on the surface in the English Channel near the Lizard, when she sighted the *Olympic*, bound up-Channel with 5,000 American troops on board. It was the kind of encounter that U-boat captains dreamed about, but on this occasion the liner's skipper, Albert Haddock, was more than equal to the challenge. Handling his ship like a destroyer, Haddock made for the unfortunate submarine at 22 knots and struck her a glancing blow. As the German vessel slid past the liner's stern, the huge wing propeller of the *Olympic* sliced into her hull like a gigantic can opener.

By the middle of 1917, the Allies were well on the way to winning the war at sea.

The intervention of America meant that ninety German liners and merchantmen lying in American ports were taken over and employed against their former owners. As we have seen, the German liners that were fit for service became troop-ships and joined their erstwhile British rivals in ferrying two million Americans 'over there'. As an example of the military capacity of these ships, on trooping service during the war, the *Olympic* steamed 184,000 miles and consumed 347,000 tons of coal. In all, she carried 41,000 civilians, 24,000 Canadian and 42,000 American troops.

When the war ended in November 1918, the Allies and an exhausted Germany had, between them, lost twelve and a half million tons of merchant shipping. Two thousand seven hundred and forty-

nine British ships alone, totalling seven and three-quarter million tons and 14,287 lives had been lost, or nearly 40 per cent of her pre-war fleet.

After the grand tragedy of 1914–18, it is incredible that there were people anywhere who sought a repeat performance. This second war caught most of the liner companies unprepared, in spite of a British Admiralty warning on 26 August 1939 that all British ships should avoid the usual trade routes. Thus, when hostilities began with the attack on Poland on 1 September, the immediate reaction of most major steamship companies was to order their Western Ocean liners to make for port and stay there.

Eight hours after Britain's formal declaration of war on 3 September, the world received a sterner declaration of

THE HOSPITAL SHIPS

A profile of the *Aquitania* as she appeared on hospital ship duties in 1915. Right: the *Britannic*, sister to the *Titanic* and *Olympic*, never served as a liner, but began life as a hospital ship during the Gallipoli campaign and was eventually sunk by a mine in the Aegean, fortunately with minimal loss of life.

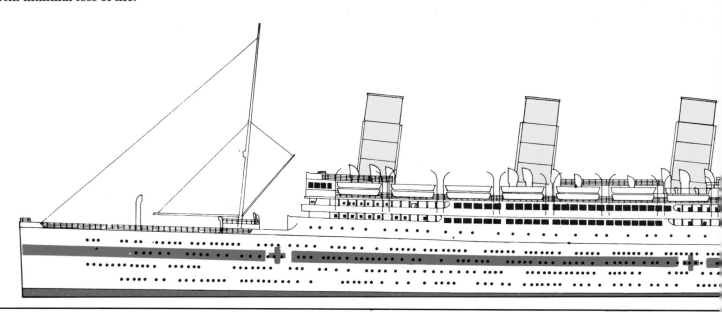

A short diversion from the cares of war: a concert programme from the Cunarder Ivernia *might well have diverted crew and passengers alike from thoughts of lurking U-boats.* Ivernia *was one of five big Cunard liners lost in the First World War.*

the kind of war that would be fought at sea during the coming years.

The Anchor-Donaldson liner *Athenia* (13,850 tons) was one of a number of intermediate vessels operating on the north Atlantic between the wars. She had left Glasgow at noon on 1 September, and, after calls at Liverpool and Belfast, headed west for Montreal, unarmed and unescorted. She carried 1,418 passengers, 300 of them American nationals. The *Athenia* was 250 miles north-west of Ireland, steaming at 16 knots, when she was struck by two torpedoes at 1945 hours local time. The liner went down in twenty minutes and 112 people died, including several Americans. Once again public opinion in both Britain and America condemned Germany, despite vigorous denials by Dr. Goebbels that a German

ship was responsible. Only after the war did the truth of the affair become known.

At the Nuremberg Trials, Admiral Doenitz, who commanded the U-boat fleet, was able to prove that he had instructed all submarine commanders not to attack unarmed passenger vessels in any circumstances. Nevertheless, the *Athenia* was torpedoed by U.30 (Kapitän-leutnant Fritz Lempke), but due to radio silence, Doenitz was not informed until U.30 returned from patrol over two weeks later. Lempke was killed later in the war, but he alleged in his report that, as the *Athenia* was steaming with all lights blacked out, he took her for an armed merchant cruiser and acted accordingly. Whatever the truth of the matter, British propagandists made great capital out of the story, particularly in America, and German interests suffered accordingly.

As the cold waters of the Western Approaches closed over the sinking hull of the *Athenia*, other liners were making a headlong dash for safety. Cunard's *Queen Mary* was on passage to New York

223

and, after a rapid voyage to the Ambrose Light Vessel fully blacked out, she was laid up at Pier 90 on the Hudson River. Alongside her, at Pier 88, she found her giant French rival, the *Normandie*, also under orders to remain in New York indefinitely. Though none knew it at the time, the beautiful *Normandie* was never to sail again, while the *Queen* had a brilliant war career ahead of her. Cunard also kept their second (and brand new) *Mauretania* at New York, but the *Aquitania* completed a voyage into Southampton and then made another crossing.

During this dangerous undertaking, while the few Americans on board prayed fervently for the trip to pass as speedily and uneventfully as possible, it is alleged that British passengers regularly changed into dinner jackets and evening gowns as though the war did not exist. They then proceeded to gather in the ship's splendid Palladian Lounge and make small talk about

cricket and the weather, while expecting a torpedo at any minute! The *Aquitania* carried on until her requisition as a troopship in November, but in the meantime she repatriated many of the American citizens caught · up in the European emergency. With practically all the French, British and German liners out of service, hundreds of Americans were left stranded at Channel ports without return reservations. Many of these were reluctant to commit themselves to a belligerent vessel, and they were eventually rescued by the *Manhattan, Washington* and *President Roosevelt*, carrying the American flag prominently painted on each side, together with the ship's name.

Another superliner away from her home port in those autumn days of 1939 was the North German Lloyd Blue Riband challenger *Bremen*. Her sister *Europa* lay safe in Bremerhaven, but the *Bremen* was in New York. In the small hours of 1 September she slipped from

her berth and, without passengers, sailed down the Hudson blacked out and homeward-bound. The Atlantic Ocean is nothing if not big, and naval patrol aircraft in 1939 did not possess the range that they do today. To the world in general, and the Royal Navy in particular, the *Bremen* just vanished. In fact, Kapitän Ahrens had taken her north into waters where no Atlantic liner would venture in peacetime (the war would in due course send other superliners to strange seas) and, six days out from New York, she turned up in Murmansk, naval base of Germany's new-found ally, the Soviet Union, and the only ice-free port in winter on the north Russian seaboard.

There she stayed, like some huge trapped whale, until the second week of December 1939. Then, in atrocious weather for which he had waited for weeks, Ahrens made his final dash to Germany. Incredible though it was, Royal Navy submarine commanders were still

*Below: victim of the
R.A.F. or a saboteur?
The great* Bremen
*heels to starboard as
fire sweeps through
her on 19 March 1941.
Bottom: after her
dramatic escape from
the Arctic, the*
Bremen *was refitted
as a troopship and the
Kriegsmarine planned
to use her in
Operation Sea Lion,
the proposed invasion
of England. At this
time the* Bremen *was
dazzle painted.*

under orders to stop and warn all unarmed merchant vessels before sinking them. So when H.M. Submarine *Salmon* sighted the *Bremen* making 25 knots in the Skaggerak on the early morning of 11 December, Lt. Cdr. Bickford brought his boat to the surface and promptly ordered the liner to stop. But signals from the *Bremen* brought Luftwaffe patrols immediately to the spot and Bickford was forced to submerge, so losing his huge prize. The liner arrived safely in Bremerhaven the same evening. There, she and the *Europa* remained laid up, and neither liner played any part in the war. The *Bremen* was gutted by fire during an air raid on the night of 18/19 March 1941, but it remains a matter of doubt whether the *Bremen* was a victim of the Royal Air Force or arson by a disgruntled crew member. Scuttled in an attempt to quench the fires, she lay on the bottom of her Bremerhaven berth until, declared beyond repair, she was broken up for scrap.

Germany's third largest ship, the North German Lloyd Columbus *(32,581 tons) burning furiously after her crew had set her on fire off the coast of the United States to avoid capture by a British destroyer.*

The *Europa* did a number of trooping voyages after the war, returning American troops, before being handed over to the French Line to become its very popular *Liberté* in 1946.

One other large German liner was at sea when war broke out. This was Germany's third largest ship, the North German Lloyd liner *Columbus* (32,581 tons), engaged in the cruising trade from New York to the Caribbean. A few days after the *Bremen* reached safety, the *Columbus* made her own bid to return home. This time fortune was with the British. They caught the *Columbus* 300 miles off the coast of Virginia and the Germans opened the sea-cocks and let their ship sink under them to deny her use to their enemies.

In March 1940, the *Normandie* and the *Queen Mary* were joined at New York by the new Cunarder *Queen Elizabeth* which had sailed secretly from the Clyde in a semi-finished condition. The huge vessel was an obvious target for the Luftwaffe if she remained for any length of time in her Glasgow fitting-out berth, and the Admiralty wanted her out of the way so that warships could take her place in the shipyard. With a crew recruited from the *Aquitania*, who thought they were bound only for Southampton, the world's largest ship slipped out of the Clyde and, under the command of Captain John Townley, made straight for New York without any regular trials of her machinery.

So for the first – and last – time, the three largest passenger ships ever built lay alongside each other, tied up at Piers 88, 90 and 92 in New York's Hudson River. The trio remained for three weeks only. On 29 March 1940 the *Queen Mary* left for Sydney, Australia, where she was converted into a troopship, spending the remainder of the war in that role. Most of her early war service was spent on the run between Australia and the United Kingdom, but at the end of 1941 she transferred to the Atlantic, sailing between New York, Halifax and the Clyde. The high speed of the two *Queens* (the *Queen Elizabeth* was also converted at Sydney)

allowed them to sail unescorted, relying on their speed to carry them safely past the attentions of any marauding U-boat or surface raider. Many are the tales of the submarine skipper who watched the *Queen Mary* sail slap across his bows 'just after I had used my last torpedo!'

Between them, the two *Queens* lifted a total of 1,577,000 service personnel during their war service and did so without losing a single life due to enemy action. Sir Winston Churchill once estimated that the two ships had helped to shorten the war in Europe by at least a year and he continued: 'To those who brought these two great ships into existence, the world owes a debt that it will not be easy to measure.'

The *Normandie* languished in New York, tended only by a skeleton crew and costing her owners $1,000 per day in port charges, for the first two years of the war. The crew did their best, but fought a losing battle against the overall deterioration which sets in when a ship is laid up. Then, when America entered the war in December 1941, the *Normandie* was taken over by the United States Navy and her owners were promised compensation for their loss. There followed a sequence of events which ended in the destruction of possibly the finest ship ever to sail the Atlantic.

The tricolor was hauled down for the last time on 16 December 1941 and the S.S. *Normandie* became the U.S.S. *Lafayette*, named after the great eighteenth-century Franco-American statesman. It was the last happy note in the whole *Normandie* affair. The contract for her conversion into a trooper was won by the Robins Dry Dock and Repair Co. Inc. who planned to carry out the work where the *Normandie* lay at Pier 88. The contractors were up against a tight programme as the ship was scheduled to sail from New York by 14 February 1942. Matters were not improved by continually changing, contradictory instructions from the Navy Department, and it is not surprising that fire precautions on board were somewhat overlooked. Part of the

Above: the tremendous troop-carrying capacity of the two Queens *is clearly demonstrated in this photograph of the* Queen Elizabeth *with 15,000 American and Canadian troops on board.*
Left: popular in peace and war: a souvenir of Canadian regiments which travelled home in the Ile de France.

conversion work involved removing pillars from the ship's many lounges using flame cutters, and on 9 February while the work was at its height, sparks from an oxy-acetylene torch ignited a pile of kapoc life jackets, which had – with incredible carelessness – been stored within arm's length of the flame cutters.

Due to the run-down state of the ship, there was no effective fire-fighting equipment to hand and a serious fire took hold. The New York Fire Brigade rushed forty-three fire-fighting appliances to the pierside and the battle to save the ship was on, as a pall of smoke began to drift over Manhattan. But the fire had taken hold and the brigade never really had a chance. The ship was soon without power and the best that could be done was to evacuate the 3,000 people working on board, which was carried out with the loss of only one life. That apart, the brigade pumped tons of water aboard with the inevitable result. Twelve hours after the fire started, the great *Normandie*, the pride of France and dream ship of millions, heeled over to port and lay a burnt-out wreck on the bottom of New York harbour. She had not died by the hand of an enemy, but the pressures of war had destroyed her as surely as any U-boat.

The hulk was raised sufficiently by October 1943 for it to be towed away, and on the second day of the month, the *Normandie* finally left Pier 88. She had been there for over four years. In 1946, the United States Navy sold her for scrap for a paltry $161,680: she had cost $48 millions!

The other big liners carried on trooping; the *Aquitania* (doing it all for the second time), the *Ile de France*, the *Mauretania*, the *Nieuw Amsterdam*, the *America*, the *Manhattan* and the *Washington* all played their part and all survived. But others did not, and three of the largest all fell victim to the superliner's deadliest enemy in peace or war – the airplane.

Among the earliest wartime losses was the Canadian Pacific flagship *Empress of*

231

Britain. On 26 October 1940, while on passage from Capetown to Liverpool with 643 passengers, she was attacked and set on fire by a Luftwaffe bomber 60 miles north-west of Donegal. The crippled liner was taken in tow but thirty-six hours later was torpedoed and sunk by U.32.

Both the big Italian liners went down under bombs from Allied aerial attacks. The *Conte di Savoia* was the first to go. After some sporadic trooping, she was laid up in Venice, where American bombers sank her on 11 September 1943. The *Rex* outlived her sister by just one year. On 7 September 1944, aircraft from the Royal Air Force caught her under tow off Cape d'Istria near Trieste and attacked her at intervals over the next forty-eight hours until the great ship lay capsized and burning.

Earlier in this chapter we discussed the losses of armed merchant cruisers during the First World War. The experience proved that passenger liners did not make good cruisers, but that did not prevent the Royal Navy requisitioning fifty-six liners, all twin-screw vessels with a service speed of 15 knots or more, and putting them into service in 1939–40. But they were so vulnerable to attack that fourteen had been lost by the summer of 1941 and after that time they were phased out of active operations. But before they went, two of their number fought actions that for sheer courage must be reckoned among the most heroic naval actions of all time.

Armed liners had been used with some success in enforcing the blockade on trade with Germany in the waters around Iceland in the First World War and they were employed in the same fashion in 1939. On 23 November of that year the P. & O. liner *Rawalpindi*, 16,600 tons, mounting six 6 inch guns, encountered the German battlecruiser *Scharnhorst* while on blockade duties off the southeast coast of Iceland. Captain Kennedy, aware that the *Scharnhorst* carried a main armament of nine 11 inch guns and a host of smaller weapons, veered away to port only to find the way barred by the battle-

cruiser's twin sister, the *Gneisenau*. There was no alternative but to fight it out and the *Rawalpindi* was first to open fire. The fight was hopeless from the start and soon the British ship was ablaze from end to end, although some brave spirits were still manning her guns when she blew up after the slaughter had lasted an hour. Kennedy and all his crew save eleven perished with the ship.

Another far more dangerous use for the armed merchant cruisers was that of convoy escort. In the early part of the war, the Kriegsmarine sent a number of its heavy ships raiding in the Atlantic and Indian oceans, and there was always the chance of a convoy being attacked. Quite often the only available escort was an armed merchantman and this was so when Convoy HX 84 came under attack on 5 November 1940. The escort was provided by the Aberdeen and Commonwealth Line's *Jervis Bay* (13,850 tons) which carried eight 6 inch guns and was under the command of Captain Fogarty Fegen, R.N. There were thirty-seven ships in the convoy and about mid-afternoon a heavy warship came in sight whose identity was not long in doubt. Ordering the convoy to scatter, Fegen turned his improvised cruiser towards the warship, sending as he did so the following signal: 'From *Jervis Bay* – Immediate to Admiralty – Halifax Convoy HX.84 attacked in position 52.26 North, 32.34 West by *Admiral Scheer*. I am attacking.'

The *Jervis Bay* had been built for the Australian migrant trade and she had no chance at all in a fight with the *Scheer*, a 12,000-ton pocket battleship with six 11 inch guns, whose diesel motors pushed her along at 26 knots. Fogarty Fegen and every man of his crew knew what lay ahead, but without hesitation the *Jervis Bay* made straight for the *Scheer* and opened fire. This move allowed the bulk of the convoy to escape as Kapitän Krancke of the *Scheer* pounded his valiant opponent into a blazing ruin. He took but half an hour to do so, but that was long enough. Fegen died with his ship, but after darkness fell, the Swedish

WILHELM GUSTLOFF

Hitler attended the launching of this prestige project for the Third Reich, but the liner's career was to end in tragedy, as an even greater tragedy engulfed Germany. In January 1945, the liner was en route between Gdynia and Kiel with more than 6,000 refugees on board, when it was struck by a Russian torpedo during one night. In the darkness and high seas a rescue fleet was only able to save 500 people. This was the greatest shipping disaster ever and it is not known exactly how many people perished.

SOS - SOS - WILHELM GUSTLOFF - STOP -
SIND SCHWER GETROFFEN-STOP- SCHIFF
SINKT MIT 6000 MENSCHEN-SOS-SOS

Vor der Katastrophe in der Nacht vom 30. zum

Um 4 Uhr früh kamen die ersten verwundeten Soldaten. Dann ergoß sich Flüchtlingsstrom auf F

Ein Augenzeuge berichtet über die furchtbarste Schiffskata
5000 Menschen in der Ostsee ertranken. V

Was bringt die
N.-S.-Gemeinschaft

Kraft durch
Freude

Reisen, Wandern und Urlaub
Siedlung und Selbsthilfe
Volkstum und Heimat
Sport, • Ausbildung
Schönheit der Arbeit
N.S. Kulturgemeinde

freighter *Stureholm* returned to the scene of the action and picked up sixty-five survivors from the *Jervis Bay*. For Fogarty Fegen there was a posthumous Victoria Cross and for the *Jervis Bay* a permanent place on the roll of honour of historic ships.

By 1942, the war at sea in the west had become almost exclusively a submarine campaign, while in the eastern seas the aircraft carrier emerged as the most powerful naval weapon of both the United States and the Japanese Empire. For the liners, the only meaningful role was trooping and most of them continued in that service until the end of the war. Their contribution to the war effort of both sides can be measured by the size of the casualty lists when one of them was sunk, fortunately an infrequent occurrence. The grisly record for the highest loss of life in any sea disaster was achieved during the sinking of Cunard's *Lancastria* on 17 June 1940 by Luftwaffe dive-bombers off Saint Nazaire during the British withdrawal from France. Of 6,000

The damaged bows of the Queen Mary *after her collision with the* Curaçao.

The British light cruiser Curaçao *was sliced in half by the giant bow of the* Queen Mary, *due to an error of navigation on 2 October 1942. Three hundred and twenty-nine members of the* Curaçao's *crew were lost in the accident.*

troops on board, over half perished. Three bombs exploded inside the ship and she foundered in less than twenty minutes. The exact number that died has never been established. But this uncertain total was exceeded in January 1945 when the German liner *Wilhelm Gustloff*, en route between Gotenhafen (Gdynia) and Kiel with more than 6,000 refugees from East Prussia, was torpedoed by the Soviet submarine S13 (Lt. Ivan Marinesko). The result was the greatest ever disaster at sea. No one knows with any certainty the total number of people aboard the *Gustloff*, but in darkness and high seas, the rescue fleet led by the cruiser *Admiral Hipper* were only able to save 500 people and at least 5,800 were drowned.

Another grim incident marred the excellent war record of the *Queen Mary*. On 2 October 1942, while inward bound for the Clyde with a full load of American troops, a navigational error by an escorting cruiser H.M.S. *Curaçao* led to a collision between the great Cunarder and the warship. The bow of the *Queen Mary*

sliced into the *Curaçao* at 25 knots, cutting her in half, and causing but a slight shudder to pass through the liner's 80,000-ton hull. The *Queen Mary* steamed on (to stop would be to invite a U-boat attack!) leaving the cruiser to sink with the loss of 329 of her crew.

But at last it was all over and, one by one, the liners came home from the wars. There were, however, some notable absentees from the parade of returning giants. Of the eight largest ships afloat in 1939, the *Bremen*, *Rex*, *Conte di Savoia*, *Empress of Britain* and *Normandie* had all gone. Only the *Europa* and the two *Queens* survived in a war that had seen over 23,000 British and American seamen die and 574 merchant ships sunk in the Atlantic alone.

Nevertheless the war was at an end, and all that now mattered was the rapid return to service of the liner fleets. When that long awaited day at last arrived, the big ships found a very different world awaiting them. The twilight years were about to begin.

THE GHOST SHIPS

A fitting and dignified memorial of an age gone by, the QE2 is the last true superliner.

As the clouds of war receded, the liners were released from their last military duty – ferrying home the American army with its baggage train of war-brides and thousands of tiny, new-born American citizens. One by one, the ships were refitted and returned to the sea lanes.

The *Queen Elizabeth* was ready for her commercial maiden voyage in October 1946. Stripped of war-time grey, she now emerged in Cunard colors for the first time, while an army of painters chipped and brushed away at her huge structure. Inside, the veneers and panels in 'Odeon' style, so beloved in the thirties, were installed in their somewhat dated glory. To the war-weary British, the menu provided in the liner's first-class dining room seemed gargantuan. It was faith-fully reprinted in every newspaper, where Cunard were careful to point out that such delicacies had been obtained in far away Canada and shipped over in the *Aquitania*. Her sponsor, Queen Elizabeth, went aboard the ship during acceptance trials and took her two teenage daughters along for the ride (twenty years on, one of them would give her own name to the *QE2*). Then, with every ticket sold, and Russian Foreign Ministers Molotov and Vishinsky on the passenger list, the largest ever passenger ship headed west, six years late in her intended true role. Her captain for the trip was James Bissett, who had been Rostron's second officer in the *Carpathia* during the *Titanic* rescue.

The *Queen Mary* joined her sister a year later and the long awaited weekly service using two ships was at last a reality. Cunard cashed in on the demand for Atlantic passages and these post-war years were the most profitable in the company's one-hundred-year history. In one month alone, July 1950, 146,000 Americans travelled to Europe in British ships. By 1949 Cunard had six large liners back in service, although the stout old *Aquitania* went for scrap in 1950 after twenty-eight years of service.

North German Lloyd's pre-war record-breaker *Europa* was handed over to the

Dance!
a about!

RTY R·K·O
RADIO
PICTURES
GRAINGER

One of the great favorites of the North Atlantic route was the Liberté, *the flagship of the Compagnie Générale Transatlantique. The atmosphere on board was considered to be the most frivolous and exciting of any of the great liners, a reputation which was even further enhanced when it was used as the setting for the Hollywood movie* The French Line, *starring Jane Russell.*

French Line to become its *Liberté*. She arrived at Le Havre in November 1946, and the following month broke loose from her moorings in a gale. The ship drifted on to the wrecked *Paris* and sank, for the second time in her career, incredibly again on an even keel. Raised in April 1947, it was not until 1950 that she was ready for service, but then proved very popular. Built in Germany, she nevertheless assumed a French ambiance, and her cuisine was acclaimed as the best afloat. The *Liberté* also acquired an unsought reputation as the Atlantic's floating palace of promiscuous romance, an idea that a Hollywood film, *The French Line*, starring Jane Russell and made on board, did little to dispel.

The *Liberté* sailed in company with the *Ile de France*, also rebuilt with her three stacks reduced to two, and the Dutch later added their *Nieuw Amsterdam*. When the *United States* came in 1952, the Atlantic ferry began to acquire something like its pre-war glory.

Appearances can be deceptive, however, and no more so than on the Atlantic in the 1950s. Whereas eighty-six express

The elegant, soaring bow of the Nieuw Amsterdam II *(1938) is picked out in this detail from a poster advertising the Holland America Line. This liner started life as a troopship during the Second World War, and only later took up its civilian role at the end of the war.*

liners worked the service in 1939, by 1953 the figure was down to forty, and some of these were over forty-years-old. In that year 38 per cent of travellers preferred to cross by air. Four years later the total had risen to 55 per cent. By 1960, the airlines had taken 69 per cent of the business, carrying almost two million passengers. By then, it was clear to all who cared to notice that the big express liners were doomed as commuter vehicles.

The airplane killed the superliners. Other influences played a part, but the outcome had never been in doubt from the very moment the Wright brothers first flew their rickety but practical aircraft, four years before the *Mauretania*'s maiden voyage.

Paradoxically, it was not an airplane that first challenged the liners from the skies, but ships that rivalled them in size – the great German dirigibles *Graf Zeppelin* and *Hindenburg*. Dr. Hugo Eckener, creator of the *Graf*, had proved by 1936 that his big airship was a practical passenger carrier with good earnings potential. That year he introduced the larger *Hindenburg* and together the two airships operated across the Atlantic to Rio or New York throughout the summer. But Eckener's dream of worldwide scheduled services was destroyed when the *Hindenburg* blew up at Lakehurst, New Jersey, in May 1937 as she was ending her first crossing of the new season.

The shipping companies breathed again, but even so, in the years 1934 to 1938, the *Mauretania, Olympic, Homeric, Berengaria* and *Majestic* were all taken out of service, and replaced by just one ship – the *Queen Mary*. Then, just two months after the blazing end of the *Hindenburg*, the challenge was renewed when, on Independence Day, 1937, Pan American's *Clipper III* and Imperial Airways *Caledonia* lifted off on either side of the Atlantic to start a weekly service by flying boat. It was, all in all, a modest enough start, but aviation, of all sciences, was boosted by the war years far beyond anything engineers thought possible in 1939.

Aircraft grew in size – they had to be capable of carrying the large bomb loads demanded by air staffs – and their operating range increased accordingly. Britain produced bombers capable of carrying ten tons to Berlin, and the American B29s ranged the far reaches of the lonely Pacific to rain death by the ton on the Japanese home islands. Late in the war, a major break-through in engineering technology occurred when both Britain and Germany produced gas turbines which tapped huge reserves of power and made possible the modern jet liner.

When the war ended, the big long-range airplane was a fact. With jet power installed, man had discovered the perfect alternative to the superliner as a method of intercontinental passenger transport. Britain produced the world's first jet liner in passenger service, the de Havilland Comet, as early as 1952, but technical problems with the aircraft's structure led to early crashes, and the lead was lost to the giant American Boeing company who put their B707 into service in 1958. This famous airplane could lift up to 150 passengers in comfort and get them across the Atlantic in eight hours or so. The shipping companies had no possible answer to this kind of opposition and they knew it. The facts spoke for themselves. In 1960 alone, aircraft made 70,000 Atlantic crossings.

By the early 1960s, liners were being withdrawn from service at the rate of ten a year or more. But the companies did not give up without a fight and several new ships appeared when commercial success appeared unlikely. North German Lloyd came back on service with the fourth *Bremen*, and Holland America's *Rotterdam*, revolutionary in appearance and design, appeared in 1959. The lovely *France* (1961) of the Compagnie Générale Transatlantique rivalled the *Normandie* in taste and style, and as late as 1966, Italia's *Michelangelo* and *Raffaello* advanced standards of passenger comfort to new high levels.

On other seas, where emigrant traffic to Australia was in demand and the fast mail run to South Africa carried out by

passenger vessels, big liners could still be profitable. The *Canberra*, *Oriana* and *Windsor Castle* were all products of the sixties and all three were the largest ships laid down for their respective companies.

On the Atlantic, though, the airplane reigned supreme. From 1958 onwards, the big jets thundered through the troposphere, carrying ever increasing numbers of passengers who gladly accepted travelling conditions more cramped than any steerage cabin in exchange for a passage time of ten hours or less. In these years, the big liners became the ghost ships of the Atlantic and anyone who crossed in winter found himself almost alone in the vast lounges and saloons, while the hired help outnumbered the paying customers by three or four to one!

This state of affairs forced the liner companies to look elsewhere for revenue. There was only one possible answer – cruising – and only one country whose prosperity could provide the necessary market – the United States. So every winter, when the summer season on the Atlantic was over, the big ships headed south for Florida, Bermuda and the warm, gentle blue seas of the Caribbean. For some, the venture bought new life and profit. The *Rotterdam* had been designed for off-season cruising and her winter round-the-world cruise has become a regular feature of the cruising trade.

For others, the going was anything but easy. In the thirties, Cunard had made money with cruises by the *Mauretania*, so the *Queens* were sent out from New York to repeat the process. It was asking too much of ships designed to stand the worst that north Atlantic winters could throw at them. Their size limited the number of ports of call, and Caribbean heat turned passenger accommodation into involuntary saunas. The *France* and the *United States* fared better, being fully air-conditioned, but they were big ships to fill and the port problem was as bad for them as the *Queens*. After a few seasons, the losses were too great to bear any longer, and the *Queens* and the *United States* all came out of service between 1967 and 1969.

Nevertheless, cruising did provide a lifeline for passenger liners as a breed, and several purpose-designed cruising vessels were built in the late sixties and early seventies. The largest was Home Lines *Oceanic*, a 39,000-ton beauty from the same Trieste yard that produced the *Conte de Savoia* and the *Raffaello*. She possesses a huge glass-covered lido amidships and has enjoyed great success in the American cruise market. Most companies however have gone for medium-size ships of an average of 15,000 tons, allowing accommodation for about 500 passengers in great luxury. Fully air-conditioned outside cabins with *en suite* facilities are essential for success. It is all a far cry from Charles Dickens and his hard mattress and chamber pot on the *Britannia* 130 years ago. Typical of these modern cruise liners are the Norwegian *Royal Viking* sisters and the *Song of Norway*, and Cunard's *Cunard Ambassador*. Although a summer cruise season is operated in European waters, the main bases for these ships are now in the Caribbean and on the west coast of the United States, where operations are more profitable, especially after the world oil crisis of 1973.

By the mid-seventies, the last scheduled passenger services had been withdrawn and the day of the liner was over. But one house flag still flies on the Atlantic from the mast of a superliner. By one of the happier accidents of history, the

245

QUEEN ELIZABETH 2
The last of the line

There is very little chance that a liner as large as the *QE2* will ever be built again. Born in the midst of economic difficulties, both for her owners and builders, the last of the superliners was also delayed by technical trouble before she finally entered into her true role as flagship of the great Cunard Line on the north Atlantic run. Much of her income now comes from luxury round-the-world cruises, but she is still the only important ship to maintain a regular passenger service between Europe and New York, thus remaining a contemporary memorial to an age gone by.

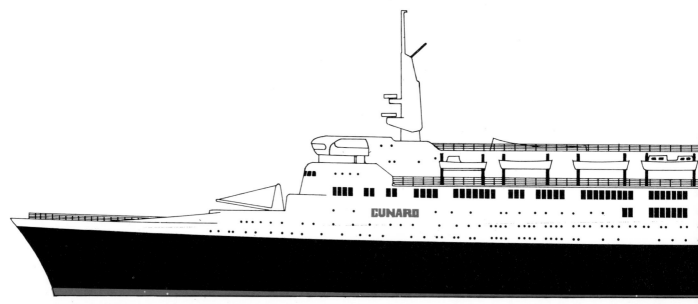

5d RMS Queen Elizabeth 2

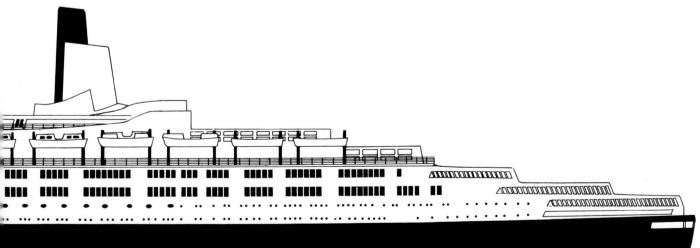

A Royal Marine bomb disposal squad was parachuted into the Atlantic alongside the QE2 on 18 May 1972 after a warning that a bomb had been hidden on board. The high drama of the incident was short-lived : no bomb was found.

Cunard Line is the last big company to offer a steamship service between Europe and New York, just as it was the first to pioneer the route long ago in 1840. Every spring, as the winter cruise season ends, Cunard's flagship *Queen Elizabeth 2* sails west from Southampton and Le Havre on the old great circle route to New York and maintains the ferry until the autumn, when she returns to the Caribbean.

Originally conceived as a replacement for the ageing *Queen Mary*, the ship was first planned as early as 1960 under the title *Q3*. Cunard then envisaged a 70,000-ton-plus monster, designed only for the north Atlantic run, and, once again, government aid was sought. Fortunately for Cunard and the British taxpayer, the idea was dropped in favour of a more flexible design that came to be known as *Q4*. This was for a 60,000-ton ship, with a capability to pass through both the Suez and Panama Canals, able to deal with Atlantic winter gales and the heat of a Caribbean summer and so combine an Atlantic ferry role with world-wide cruising possibilities. Traditional layouts familiar to veteran Atlantic travellers were scrapped, including the hallowed enclosed promenade, and Britain's top designers produced some of the best public rooms ever installed in a liner.

Q4 was ready for launching on 20 September 1967 and Queen Elizabeth II travelled to John Brown's Clydebank yard, just as her grandmother and mother had done, to name the new ship. Cunard showed the same reticence in announcing the chosen name as they had done thirty-four years earlier with the *Queen Mary*, and it was not until the actual words of the Queen, 'I name this ship *Queen Elizabeth the Second*', that the secret came out.

The *QE2* was born against a background of economic trouble for both Cunard and John Brown. By the time the ship left the builder's yard in November 1968, John Brown had been absorbed into Upper Clyde Shipbuilders and the demise of that consortium was but three years ahead. Then technical trouble hit

the new liner. First, oil was allowed to contaminate boiler feed water. But far worse was to follow. Turbine vibration led to a refusal by Cunard to accept the ship and repairs led to a five-month delay on her maiden sailing to New York. She got away at last on 2 May 1969 under the command of Captain Bill Warwick: the *QE2* entered into her rightful kingdom, the High Seas.

Ranging the wide Atlantic and far beyond, she has, in the succeeding years, fulfilled expectations beyond most people's belief and operated consistently at a profit, aided by some shrewd management from Trafalgar House, who bought the Cunard Line in 1971. In doing so, she has shared the excitements and dangers of the seventies. The I.R.A. have

attempted to smuggle arms for Ulster in the new superliner and she has twice been involved in the Middle East crisis. On 18 May 1972 a telephone call in New York claimed that the liner, then in mid-ocean, carried a hidden bomb and the following day a Royal Marine bomb disposal squad parachuted into the sea alongside. The drama ended in anti-climax, as nothing was found. Then, in the summer of 1973, the *QE2* visited Haifa on a charter trip to celebrate the first twenty-five years of the State of Israel. Cunard came under verbal fire from the Arab States, but it was not until July 1974 that President Sadat of Egypt claimed he had foiled Palestine guerrilla plans to hijack an Egyptian submarine and destroy the *QE2* at sea!

It is unlikely that so large a passenger liner will ever be built again, unless some distant energy crisis forces man to return again to simple fuel like coal or wood. The *QE2* may well be the last of her breed, but she is a true superliner. While she remains, she is not only a living ship, but also a super-memorial to the days and ships that are gone, and to an age that measured greatness by style and wealth. Other memories are there too. Her hull creaks in a Force 8, one can still get morning bouillon on deck, the wind sounds in the rigging and the siren booms out at noon. Below, the gentle hum of the engines mingles with the casual chat at dining tables, as the ship's world gently passes into night, just as the liners themselves have passed away.

Index

Figures in italics refer to illustrations

Bibliography

ANDERSON, R.: *White Star*, Prescot, 1964

ARMSTRONG, W.: *Atlantic Highway*, New York, 1962

BEESLEY, L.: *Loss of the S.S. Titanic*, Boston, 1912

BONSER, N.: *North Atlantic Seaway*, Prescot, 1955

BRINNIN, J. M.: *The Sway of the Grand Saloon*, New York, 1971

CORSON, F. R.: *The Atlantic Ferry in the 20th Century*, London, 1930

CRONICAN, F. and MUELLER, E. A.: *The Stateliest Ship*, New York, 1971

DIGGLE, CAPT. E.: *The Romance of a Modern Liner*, London, 1930

DODMAN, F. E.: *Ships of the Cunard Line*, Southampton, 1955

DUNN, L.: *Passenger Liners*, Southampton, 1961

DUNN, L.: *Famous Liners of the Past*, Southampton, 1964

LE FLEMING, H.: *Cunard White Star Liners of the 1930s*, London, no date

GIBBS, C. R. V.: *British Passenger Liners of the Five Oceans*, New York, 1963

HUGHES, T.: *The Blue Riband of the Atlantic*, London, 1974

ISHERWOOD, J. H.: *Steamers of the Past*, Liverpool, 1966

LACEY, R.: *The Queens of the North Atlantic*, London, 1973

LEE, C. E.: *The Blue Riband of the Atlantic*, London, 1931

Lloyds Register of Shipping (various editions), London, 1897–1974

MARCUS, G.: *The Maiden Voyage*, London, 1969

MARR, G.: *The Queens and I*, Southampton, 1973

MAXTONE-GRAHAM, J.: *The Only Way to Cross*, New York, 1972

MIDDLEMAS, K.: *Command the Far Seas*, London, 1961

POTTER, N. and FROST, J.: *The 'Elizabeth'*, London, 1965

POTTER, N. and FROST, J.: *The 'Mary'*, London, 1961

POTTER, N. and FROST, J.: *Queen Elizabeth 2*, London, 1969

SCHIEDROP, E. B.: *The High Seas*, London, 1939

STANFORD, D.: *Ile de France*, London, 1960

STEVENS, L.: *The Elizabeth – passage of a Queen*, London, 1969

TALBOT, F. A.: *Steamship Conquest of the World*, Philadelphia, 1912

TALBOT-BOOTH, E. C.: *Merchant Ships*, London, 1943

TALBOT-BOOTH, E. C.: *Merchant Ships*, London, 1949

TALBOT-BOOTH, E. C.: *Merchant Ships*, Liverpool, 1959

TALBOT-BOOTH, E. C.: *Ships of the British Merchant Navy*, London, 1932

TALBOT-BOOTH, E. C.: *Ships and the Sea*, London, 1943

WINCHESTER, C. (ed.): *The Queen Elizabeth*, London, 1947

WINCHESTER, C. (ed): *Shipping Wonders of the World*, London, 1937

Acknowledgements

This book could not have been produced without the help and cooperation of a great many individuals and organizations. In particular, the author would like to thank the staff of the City Reference Library in Bristol, England, for their help in tracing information in many early publications and newspapers, Mr. Ernest Dumbleton for his help and advice, and Miss Wendy Watkins, who typed the manuscript and suffered many alterations to it with great tolerance and good humor.

The author and publishers would also like to extend their thanks to the following individuals for their great help in obtaining information and illustrations: Berty Bakker (Meijer Pers, Amsterdam), W. Beazley (Canadian Pacific, London), Commander Blake (*Great Britain* Project, Bristol), Jeff Blinn (Moran Towage, New York), René Bouvard (Compagnie Générale Transatlantique, Paris), T. Cramer (Port of Le Havre, Le Havre), G. Farnsworth (*Cork Examiner*), J. Finler, Geraldine Finney, Jane Foster (Union Castle, London), J. O'Neil (Cunard Archive of Liverpool University) and N. Thomas (City Museum, Bristol).

The author and publishers are particularly grateful to the following shipping and shipbuilding companies, institutions, publications and agencies who have supplied original illustration material from their own archives and collections and have given permission for its use in the present book: Ansaldo, Genoa; Associated Press, London; Blohm und Voss, Hamburg; Bristol Corporation, City Museum; British Transport Docks Board; Brunel University, England; Cammell Laird, Birkenhead; Canadian Pacific Steamship Co., London; Chantiers de l'Atlantique, Saint Nazaire; Compagnie Générale Transatlantique, Paris; Cunard Line, London; *Cork Examiner*; City of Liverpool Libraries; City of Liverpool Museums; City of Long Beach Museum of the Sea; *Daily Mirror*, London; German Government Press Service, Bonn; Hapag Lloyd, Hamburg; Harland & Wolff, Belfast; Holland America Line, Rotterdam; Hong Kong Government; Italcantieri, Trieste; Italia Line, Genoa; Mansell Collection, London; Ministry of Defence (Navy), London and Bath; Moran Towing and Transportation Co., New York; Musée de la Marine, Paris; National Maritime Museum, Greenwich; P. & O. Lines, London; Popperphoto, London; Port of Le Havre authority, Le Havre; Rotterdam Dry Dock Co., Rotterdam; Science Museum, London; Southern Newspapers, Southampton; Stewart Bale Ltd., Liverpool; Swan Hunter Ltd., Wallsend-on-Tyne; Union Castle Ltd., London; United States Lines, New York; United States Information Service, Washington; Upper Clyde Shipbuilders, Glasgow; Wilton Fijenoord, Rotterdam.